Elec

Herding
101

ALL RIGHT.
KEEP IT MOVIN'
AND DON'T GIVE ME
NO STATIC.

**Written and Illustrated
by B. K. Hixson**

Electron Herding 101

Copyright © 2002
First Printing • September 2002
B. K. Hixson

Published by Loose in the Lab, Inc.
9462 South 560 West
Sandy, Utah 84070

www.looseinthelab.com

*Library of Congress
Cataloging-in-Publication Data
Available Upon Request*

Printed in the United States of America
The juice is loose!

Dedication

For Bugg
(Courtney Jeanne Elizabeth Hixson)

Kidlette number three in the Hixson clan who was born with shpilkees in her tuchus and runs on an endless supply of energy that must be stolen from the universe via some method known only to her and Nicolai Tesla.

Mom and I look forward to many, many, many wonderful performances, entertaining conversations, deliriously silly faces, your streaming dialogue, and more mischief than a wagon load of leprechauns. What a treat it will be to grow up with you.

Acknowledgments

Getting a book out for public consumption is far from a one-man job. There are lots of thank-yous to be doodled out and at the risk of leaving someone out, we attempt to do that on this page. In terms of my physics education, at the top of the list is Mr. Ed Goffard, my fifth-grade teacher at Shaver Elementary School in Portland, Oregon. He introduced us to the world of circuit electricity and gave us the opportunity to build motors, wire an in-class telephone system, and generally hook anything that was stationary for a period of 3 minutes or less to a transformer.

As for my educational outlook (the hands-on perspective and the use of humor in the classroom), Dr. Fox, my senior professor at Oregon State University, gets the credit for shaping my educational philosophy recognizing that even at the collegiate level we were onto something a little different. He did his very best to encourage, nurture, and support me while I was getting basket loads of opposition for being willing to swim upstream. Also several colleagues did their very best to channel my enthusiasm during those early, formative years of teaching: Dick Bishop, Dick Hinton, Dee Strange, and Linda Zimmermann. Thanks for your patience, friendship, and support.

Next up are all the folks that get to do the dirty work that make the final publication look so polished but very rarely get the credit they deserve. Next, our resident graphics guru/web-head, Kris Barton gets a nod for scanning and cleaning the artwork that you find on these pages as well as putting together the graphics that make up the cover. All of that is done so that Kathleen Hixson and Kris Barton can take turns simultaneously proofreading the text while mocking my writing skills. Only then is the manuscript handed over to Susan Moore who peruses it with her scanning electron-microscope eyes and adds hyphens, commas, capitals, and other formal genera of the grammatical world that have eluded me for decades.

Once the finished product is done, the book has to be printed so that Louisa Walker, Tracy St. Pierre, Jay Brochu and the Delta Education gang can market, ship the books, collect the money, and send us a couple of nickels. A short thank-you, for a couple of very important jobs.

Mom and Dad, as always, get the end credits. Thanks for the education, encouragement, and love. And for Kathy and the kids—Porter, Shelby, Courtney, and Aubrey—hugs and kisses.

Repro Rights

There is very little about this book that is truly formal, but at the insistence of our wise and esteemed counsel, let us declare: *No part of this book may be reproduced or utilized in any form or by any means, electronic or mechanical, including photocopying, recording, or by any information storage and retrieval system, without permission in writing from the publisher.* That would be us.

More Legal Stuff

Official disclaimer for you aspiring scientists and lab groupies. This is a hands-on science book. By the very intent of the design, you will be directed to use common, nontoxic, household items in a safe and responsible manner to avoid injury to yourself and others who are present while you are pursuing your quest for knowledge and enlightenment in the world of physics. Just make sure that you have a fire blanket handy and a wall-mounted video camera to corroborate your story.

If, for some reason, perhaps even beyond your own control, you have an affinity for disaster, we wish you well. *But we, in no way take any responsibility for any injury that is incurred to any person using the information provided in this book or for any damage to personal property or effects that are directly or indirectly a result of the suggested activities contained herein.* Translation: You're on your own. Don't make toast while you are in the bathtub and if you are holding two wires make sure they aren't plugged into anything.

Less Formal Legal Stuff

If you happen to be a home-schooler or very enthusiastic school teacher please feel free to make copies of this book for your classroom or personal family use—one copy per student, up to 35 students. If you would like to use an experiment from this book for a presentation to your faculty or school district, we would be happy to oblige. Just give us a whistle and we will send you a release for the particular lab activity you wish to use. Please contact us at the address below. Thanks.

Special Requests
Loose in the Lab, Inc.
9462 South 560 West
Sandy, Utah 84070

Table of Contents

The National Content Standards (Grades K–4)

Electrical circuits can produce light, heat, sound, and magnetic effects. Electrical circuits require a complete loop through which an electric current can pass.

The National Content Standards (Grades 5–8)

Electrical circuits provide a means of transferring electrical energy when heat, light, sound, and chemcial changes are produced.

The 12 Big Ideas About Electricity & Corresponding Labs

1. All atoms are composed of protons, neutrons, and electrons. The small, negatively charged particles, known as electrons, orbit the positively charged center of the atom.

2. Electrons can be collected by rubbing two objects together producing static electricity. The effect of this static electric charge on different materials can be demonstrated a variety of ways.

3. Static electricity can be observed using a discharge rod, electro-scope, piezoelectric rocks, and / or fluorescent lightbulbs.

4. Static electricity can be collected with an electrophorus and stored in a device called a Leyden jar. The charge can be released from the Leyden jar by contact with any conductive material.

5. A Van de Graaff generator is a machine invented to produce large amounts of static electricity over a period of time.

6. Scientists use symbols to abbreviate the different components of an electrical circuit. These symbols are used when drawing electrical designs called schematics.

Even More Contents

7. Electrical circuits require a complete loop through which an electric current can pass. Circuits can either be wired in series or parallel depending on the use.

8. Materials that allow the free movement of electrons are called conductors. Those that do not are called non-conductors or insulators.

9. Electrical circuits can produce light, heat, sound, motion, and magnetic effects. The movement in an electrical circuit is regulated by a switch.

10. Batteries are storage devices for electrons. They have a positive and a negative terminal and must be connected in an electrical loop for electrons to flow through a circuit.

11. Fuses protect circuits from overheating and destroying valuable equipment.

12. Electricity can be used for entertainment, communication, education, manufacturing, safety, and convenience.

Science Fair Projects

Who Are You ? And . . .

First of all, we may have an emergency at hand and we'll both want to cut to the chase and get the patient into the cardiac unit if necessary. So, before we go too much further, **define yourself**. Please check one and only one choice listed below and then immediately follow the directions that follow *in italics*. Thank you in advance for your cooperation.

I am holding this book because. . .

___ **A. I am a responsible, but panicked, parent.** My son / daughter / triplets (circle one) just informed me that his / her / their science fair project is due tomorrow. This is the only therapy I could afford on such short notice. Which means that if I was not holding this book, my hands would be encircling the soon-to-be-worm-bait's neck.

Directions: Can't say this is the first or the last time we heard that one. Hang in there, we can do this.

1. Quickly read the Table of Contents with the worm bait. The Big Ideas define what each section is about. Obviously, the kid is not passionate about science, or you would not be in this situation. See if you can find an idea that causes some portion of an eyelid or facial muscle to twitch.

If that does not work, we recommend narrowing the list to the following labs because they are fast, use materials that can be acquired with limited notice, and the intrinsic level of interest is generally quite high.

Lab 9 • Anti-Gravity Bubbles • page 44
Lab 14 • Foil Electroscope • page 54
Lab 18 • The Zap Pak • page 64
Lab 29 • Conductivity Tester • page 103
Lab 32 • Switches • page 113
Lab 42 • Lemons in Series • page 142
Lab 49 • Quiz Board • page 160

Electron Herding 101 • B. K. Hixson

How to Use This Book

2. Take the materials list from the lab write-up and from page 193 of the Surviving a Science Fair Project section and go shopping.

3. Assemble the materials and perform the lab at least once. Gather as much data as you can.

4. Go to page 170 and read the material. Then start on Step 1 of Preparing Your Science Fair Project. With any luck you can dodge an academic disaster.

___**B. I am worm bait.** My science fair project is due tomorrow, and there is not anything moldy in the fridge. I need a big Band-Aid, in a hurry.

Directions: Same as Option A. You can decide if and when you want to clue your folks in on your current dilemma.

___**C. I am the parent of a student who informed me that he/ she has been assigned a science fair project due in six to eight weeks.** My son / daughter has expressed an interest in science books with humorous illustrations that attempt to explain electricity and associated phenomena.

Who Are You ? And . . .

Directions: Well, you came to the right place. Give your kid these directions and stand back.

1. The first step is to read through the Table of Contents and see if anything grabs your interest. Read through several experiments, see if the science teacher has any of the more difficult materials to acquire like electrical sources, chemicals for electroplating, and electical widgets, and ask if they can be borrowed. Play with the experiments and see which one really tickles your fancy.

2. After you have found and conducted an experiment that you like, take a peek at the Science Fair Ideas and see if you would like to investigate one of those or create an idea of your own. The guidelines for those are listed on page 182 in the Surviving Your Science Fair section. You have plenty of time so you can fiddle and fool with the original experiment and its derivations several times. Work until you have an original question you want to answer and then start the process, listed on page 182. You are well on your way to an excellent grade.

___ **D. I am a responsible student and have been assigned a science fair project due in six to eight weeks.** I am interested in electricity, and despite demonstrating maturity and wisdom well beyond the scope of my peers, I too still have a sense of humor. Enlighten and entertain me.

Directions: Cool. Being teachers, we have heard reports of this kind of thing happening but usually in an obscure and hard-to-locate town several states removed. Nonetheless, congratulations.

Same as Option C. You have plenty of time and should be able to score very well. We'll keep our eyes peeled when the Nobel Prizes are announced in a couple of years.

How to Use This Book

____ E. I am a parent who home schools my child/children. We are always on the lookout for quality curriculum materials that are not only educationally sound but also kid- and teacher-friendly. I am not particularly strong in science, but I realize it is a very important topic. How is this book going to help me out?

Directions: In a lot of ways we created this book specifically for home schoolers.

1. We have taken the National Content Standards, the guidelines that are used by all public and private schools nationwide to establish their curriculum base, and listed them in the Table of Contents. You now know where you stand with respect to the national standards.

2. We then break these standards down and list the major ideas that you should want your kid to know. We call these the Big Ideas. Some people call them objectives, others call them curriculum standards, educational benchmarks, or assessment norms. Same apple, different name. The bottom line is that when your child is done studying this unit on electricity you want them not only to understand and explain each of the 12 Big Ideas listed in this book, but also, to be able to defend and argue their position based on experiential evidence that they have collected.

3. Building on the Big Ideas, we have collected and rewritten 50 hands-on science labs. Each one has been specifically selected so that it supports the Big Idea that it is correlated to. This is critical. As the kids do the science experiment, they see, smell, touch, and hear the experiment. They will store that information in several places in their brains. When it comes time to comprehend the Big Idea, the concrete hands-on experiences provide the foundation for building the Idea, which is quite often abstract. Kids who merely read about series and parallel circuits, telegraphs, or the motor effect, or who see pictures of the electrons dancing across a kid's head are trying to build abstract ideas on abstract ideas and quite often miss the mark.

Who Are You ? And . . .

For example: I can show you a recipe in a book for chocolate chip cookies and ask you to reiterate it. Or I can turn you loose in a kitchen, have you mix the ingredients, grease the pan, plop the dough on the cookie sheet, slide everything into the oven, and wait impatiently until they pop out eight minutes later. Chances are that the description given by the person who actually made the cookies is going to be much clearer because it is based on their true understanding of the process, **because it is based on experience.**

4. Once you have completed the experiment, there are a number of extension ideas under the Science Fair Extensions that allow you to spend as much or as little time on the ideas as you deem necessary.

5. A word about humor. Science is not usually known for being funny even though Bill Nye, The Science Guy, Beaker from Sesame Street, and Beakman's World do their best to mingle the two. That's all fine and dandy, but we want you to know that we incorporate humor because it is scientifically (and educationally) sound to do so. Plus it's really at the root of our personalities. Here's what we know:

When we laugh ...
a. Our pupils dilate, increasing the amount of light entering the eye.
b. Our heart rate increases, which pumps more blood to the brain.
c. Oxygen-rich blood to the brain means the brain is able to collect, process, and store more information. Big I.E.: increased comprehension.
d. Laughter relaxes muscles, which can be involuntarily tense if a student is uncomfortable or fearful of an academic topic.
e. Laughter stimulates the immune system, which will ultimately translate into overall health and fewer kids who say they are sick of science.
f. Socially, it provides an acceptable pause in the academic routine, which then gives the student time to regroup and prepare to address some of the more difficult ideas with a renewed spirit. They can study longer and focus on ideas more efficiently.
g. Laughter releases chemicals in the brain that are associated with pleasure and joy.

6. If you follow the book in the order it is written, you will be able to build ideas and concepts in a logical and sequential pattern. But that is by no means necessary. For a complete set of guidelines on our ideas on how to teach home-schooled kids science, check out our book, Why's the Cat on Fire? How to Excel at Teaching Science to Your Home-Schooled Kids.

How to Use This Book

___ **F. I am a public/private school teacher,** and this looks like an interesting book to add ideas to my classroom lesson plans.

Directions: It is, and please feel free to do so. However, while this is a great classroom resource for kids, may we also recommend two other titles Electrons Ho! *if you wish to teach electricity to fourth through sixth graders and* The Electrostatic Gerbil 'Doo and Other Fashion Faux Pas *for the K–3 range.*

These two books have teacher-preparation pages, student-response sheets or lab pages, lesson plans, bulletin board ideas, discovery center ideas, vocabulary sheets, unit pretests, unit exams, lab practical exams, and student grading sheets. Basically everything you need if you are a science nincompoop, and a couple of cool ideas if you are a seasoned veteran with an established curriculum. All of the ideas that are covered in this one book are covered much more thoroughly in the other two. They were specifically written for teachers.

___ **G. My son/daughter/grandson/niece/father-in-law** is interested in science, and this looks like fun.

Directions: Congratulations on your selection. Add a gift certificate to the local science supply store and a package of hot chocolate mix and you have the perfect rainy Saturday afternoon gig.

___ **H. My luncheon group** has been talking about the effect of magnetic fields produced in electrical wires on metabolic rates and in particular assimilation of heavy metal based chelated minerals. We were wondering if you could help us out.

Directions: Nope. Try the herbal store on the corner.

Lab Safety

Contained herein are 50 science activities to help you better understand the nature and characteristics of electricity as we currently understand these things. However, since you are on your own in this journey we thought it prudent to share some basic wisdom and experience in the safety department.

Read the Instructions

An interesting concept, especially if you are a teenager. Take a minute before you jump in and get going to read all of the instructions as well as warnings. If you do not understand something, stop and ask an adult for help.

Clean Up All Messes

Keep your lab area clean. It will make it easier to put everything away at the end and may also prevent contamination and the subsequent germination of a species of mutant tomato bug larva. You will also find that chemicals perform with more predictability if they are not poisoned with foreign molecules.

Organize

Translation: Put it back where you get it. If you need any more clarification, there is an opening at the landfill for you.

Dispose of Poisons Properly

This will not be much of a problem with labs that use, study, and organize electrons. However, if you happen to wander over into one of the many disciplines that incorporates the use of chemicals, then we would suggest that you use great caution with the materials and definitely dispose of any and all poisons properly.

Practice Good Fire Safety

If there is a fire in the room, notify an adult immediately. If an adult is not in the room and the fire is manageable, smother the outbreak with a fire blanket or use a fire extinguisher. When the fire is contained, immediately send someone to find an adult. If, for any reason, you happen to catch on fire, **REMEMBER: Stop, Drop, and Roll.** Never run; it adds oxygen to the fire, making it burn faster, and it also scares the bat guano out of the neighbors when they see the neighbor kids running down the block doing an imitation of a campfire marshmallow without the stick.

Protect Your Skin

It is a good idea to always wear protective gloves whenever you are working with chemicals. Again, this particular book does not suggest or incorporate chemicals in its lab activities very often. However, when we do, we are incorporating only safe, manageable kinds of chemicals for these labs. If you do happen to spill a chemical on your skin, notify an adult immediately and then flush the area with water for 15 minutes. It's unlikely, but if irritation develops, have your parents or another responsible adult look at it. If it appears to be of concern, contact a physician. Take any information that you have about the chemical with you.

Lab Safety

Save Your Nose Hairs

Sounds like a cause celebre LA style, but it is really good advice. To smell a chemical to identify it, hold the open container six to ten inches down and away from your nose. Make a clockwise circular motion with your hand over the opening of the container, "wafting" some of the fumes toward your nose. This will allow you to safely smell some of the fumes without exposing youself to a large dose of anything noxious. This technique may help prevent a nosebleed or your lungs from accidentally getting burned by chemicals.

Wear Goggles If Appropriate

If the lab asks you to heat or mix chemicals, be sure to wear protective eyewear. Also have an eyewash station or running water available. You never know when something is going to splatter, splash, or react unexpectedly. It is better to look like a nerd and be prepared than schedule a trip down to pick out a Seeing Eye dog. If you do happen to accidentally get chemicals in your eye, flush the area for 15 minutes. If any irritation or pain develops, immediately go see a doctor.

Lose the Comedy Routine

You should have plenty of time scheduled during your day to mess around, but science lab is not one of them. Horseplay breaks glassware, spills chemicals, and creates unnecessary messes—things that parents do not appreciate. Trust us on this one.

No Eating

Do not eat while performing a lab. Putting your food in the lab area contaminates your food and the experiment. This makes for bad science and worse indigestion. Avoid poisoning yourself and goobering up your lab ware by observing this rule.

Happy and safe experimenting!

Electron Herding 101 • B. K. Hixson

Recommended Materials Suppliers

For every lesson in this book we offer a list of materials. Many of these are very easy to acquire, and if you do not have them in your home already, you will be able to find them at the local grocery or hardware store. For more difficult items we have selected, for your convenience, a small but respectable list of suppliers who will meet your needs in a timely and economical manner. Call for a catalog or quote on the item that you are looking for, and they will be happy to give you a hand.

Loose in the Lab
9462 South 560 West
Sandy, Utah 84070
Phone 1-888-403-1189
Fax 1-801-568-9586
www.looseinthelab.com

Delta Education
80 NW Boulevard
Nashua, NH 03601
Phone 1-800-442-5444
Fax 1-800-282-9560
www.delta-ed.com

Nasco
901 Jonesville Ave.
Fort Atkinson, Wisconsin 53538
Phone 1-414-563-2446
Fax 1-920-563-8296
www.nascofa.com

Ward's Scientific
5100 W Henrietta Road
Rochester, New York 14692
Phone 800-387-7822
Fax 1-716-334-6174
www.wardsci.com

Educational Innovations
362 Main Avenue
Norwalk, Connecticut 06851
Phone 1-888-912-7474
Fax 1-203-229-0740
www.teachersource.com

Frey Scientific
100 Paragon Parkway
Mansfield, Ohio 44903
Phone 1-800-225-FREY
Fax 1-419-589-1546
www.freyscientific.com

Edmund Scientific
101 E. Gloucester Pike
Barrington, NJ 08007
Phone 1-800 728-6999
Fax 1-856-547-3292
www.edmundscientific.com

Sargent Welch Scientific Co.
911 Commerce Court
Buffalo Grove, Illinois 60089
Phone 800-727-4368
Fax 1-800-676-2540
www.sargentwelch.com

The Ideas,
Lab Activities,
& Science Fair
Extensions

Big Idea 1

All atoms are composed of protons, neutrons, and electrons. The small, negatively charged particles, known as electrons, orbit the positively charged center of the atom.

Atomic Models

The Experiment

To understand electricity we have to slice apart an atom and see how it is constructed because a lot of the things that we see are a direct function of the positive and negative charges that are inside the atom.

This lab activity will introduce you to the basic ideas behind an atom. Just for the record, there is quite a bit more to atoms than what we are going to start with here—like muons, bosons, quarks, and other assorted aunts and uncles of the atomic particle world. But, to make the basic introductions we are simply going to invite you to get to know protons, neutrons, and electrons.

Materials

2   Pins, straight
1   Marker, black
4   Styrofoam balls, 2-inch diameter
2   Copper wires, 12 inches of 22 gauge
1   Puffed rice kernel

Procedure

1. First of all atoms are very small. Hold up the straight pin and take a peek at the head. There are a variety of estimates but most of them come in around 1 billion atoms—1 billion iron atoms covering the surface of the head of the pin.

It is almost impossible to imagine that many teeny, tiny, little balls packed side by side. So rather than attempt that visual exercise, let's meet the players that make up the parts of all atoms no matter how large or small.

2. Virtually every atom is composed of three main parts. In the center of the atom we have neutrons that carry a neutral charge and protons that have a positive charge. Both of these parts to the atom are relatively large and collectively are called the nucleus of the atom.The positive charge of the proton is balanced out with a negative charge of an electron. Electrons are very small and move in predictable orbits or locations called shells around the nucleus of the atom.

3. Using the black marker put a "+" on two of the styrofoam balls and a "0" on two others. The balls with the "+" represent the positively charged protons and the "0" represents the neutrally charged neutrons.

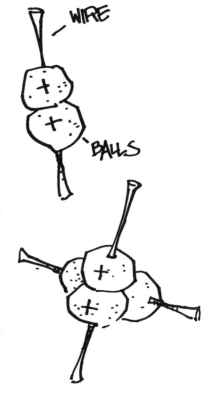

4. Thread one of the copper wires through the two proton balls and the other copper wire through the two neutron balls. Use the illustration at the right to help you out.

5. Connect the two sets of protons and neutrons together with the straight pins. When you are done, you should have something that looks like the illustration to the right.

Atomic Models

6. To complete your atomic model attach one half of the puffed rice kernel at the end of one wire and the other half on the end of the other wire. These represent the electrons.

7. Congratulations you have completed a model of the helium atom. The same atom that makes up the lighter-than-air gas we use to fill balloons and one of the simplest atoms that you will find in the Periodic Table of the Elements.

How Come, Huh?

We know that atoms are tiny—1 billion on the head of a pin is quite the pig pile. They are also made up mostly of space. Imagine a grapefruit sitting on the fifty-yard line at the center of a football field. That would be your nucleus.

Take a speck of dust and run all the way to the top of the stadium and let the speck of dust, your electron, go. The dust whirling around the stadium is representative of the distance between the electrons and the nucleus of the atom—mind boggling.

Science Fair Extensions

1. You can build more complex models of atoms. Using the Periodic Table of the Elements all you have to do is select an atom. The atomic number tells you how many protons and electrons are in the model, and if you don't get too big, you'll find the same number of neutrons. The larger the atom and the more electrons, the more electron shells you will find. If you get stuck, ask an adult to help you.

Big Idea 2

Electrons can be collected by rubbing two objects together producing static electricity. The effect of this static electric charge on different materials can be demonstrated a variety of ways.

The Neutron 'Do

The Experiment

According to our Big Idea, static electricity can be generated by rubbing two objects together. You are going to prove that idea using a rubber balloon to steal a couple of billion electrons from a pile of very fine hair and in the process create some very interesting and very funny special effects.

For the record, there are nine different static electricity experiments in this section, and they all work best when the humidity is low—or at the very least when it is not raining. If you live anywhere east of the Rocky Mountains and are reading this during the summer months, we would highly recommend skipping to the circuit electricity portion of the book—just a thought.

Materials

1   Balloon, 9 inches, round
1   Pile of hair, fine
    Low humidity, preferably

Procedure

1. Inflate the balloon and tie it off. Hold the inflated balloon next to the hair on your volunteer's head. Wave the balloon around a bit and see if you can get the hair to stick to the balloon.

2. No luck, huh? Take the balloon and rub it back and forth vigorously on the head of the person with the fine hair. After 15 to 20 rubs, gently lift the balloon off the top of your volunteer's head keeping contact with the hair shafts. Observe what happens when you move the balloon back and forth over your volunteer's head.

3. Now remove the balloon completely and take a peek at what happens to the hair. Bring the balloon back, near the volunteer's head again and observe the hair shafts. Remove the balloon again.

4. We have pretty much established the fact that hair releases electrons to a rubber balloon, which carries the charge. The lingering questions is, "Do all kinds of hair provide the same amount of electrons?"

Data & Observations

Test as many of the combinations listed below as possible and formulate an idea about which kind of hair is a good electron donor and which, if any, is not.

The first thing that you will want to do is write your prediction of what you think will happen in the space provided below. Then test 10 different folks, recording the characteristics of their hair in the data table on the next page. Judge the response by how well the balloon attracts the person's hair. Finally, write your evaluation in the chart on the next page.

Write the name of the person that you are testing in the first box. Hair color can be abbreviated as: Blonde: Bl, Black: Bk, Brown: Bn, Red: R, and White/Gray: W. Type can be abbreviated as: Coarse: C, Medium: M, Fine: F., and length can be abbreviated as: Short: S, Medium: M (shoulder length), and Long (beyond shoulders).

I predict that _____ (length), _____ (color), hair of a _____ texture will produce the most electrons when it is rubbed with a balloon.

The Neutron 'Do

	Name	Hair Color	Type	Length
1.	_____	_____	____	____
2.	_____	_____	____	____
3.	_____	_____	____	____
4.	_____	_____	____	____
5.	_____	_____	____	____
6.	_____	_____	____	____
7.	_____	_____	____	____
8.	_____	_____	____	____
9.	_____	_____	____	____
10.	_____	_____	____	____

Circle the names of the people whose hair produced lots of electrons for the balloon.

Observations: _____

_____

_____

How Come, Huh?

When we start out, everything is balanced. The balloon has the same number of electrons and protons and the charge on the hair shafts is balanced. Think back to the first lab in the book. In a balanced atom there is one electron for every proton and vice versa. Everyone is happy.

BIG NEGATIVE CHARGE

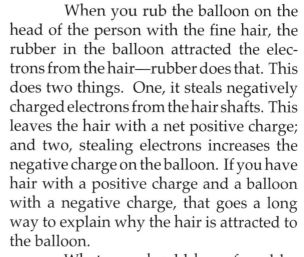

When you rub the balloon on the head of the person with the fine hair, the rubber in the balloon attracted the electrons from the hair—rubber does that. This does two things. One, it steals negatively charged electrons from the hair shafts. This leaves the hair with a net positive charge; and two, stealing electrons increases the negative charge on the balloon. If you have hair with a positive charge and a balloon with a negative charge, that goes a long way to explain why the hair is attracted to the balloon.

What you should have found by testing different kinds of hair is that fine hair works the best. Color and length tend to not have any effect on the ability of the hair to donate electrons. The one determining factor is the thickness of the hair. The reason for this is that finer hair has a greater <u>overall</u> surface area. <u>More shafts of hair in the same amount of space</u> means that more electrons are exposed to the rubber surface of the balloon where they can be stolen.

Science Fair Extensions

2. Once you have experimented on human hair, why not branch out to other animals? Nab the neighbor's cat and see if a charge can be generated there. Try your dog, rabbit, gerbil, or pet llama.

3. There are also skins available; fox, coyote, beaver, mink, deer, and badger are some of the more common furs available. However, all roadkill is off-limits for experimenting.

Stuck-Up Balloons

The Experiment

Now that you have given everyone in the family a new 'do, we are going to extend the previous experiment. We'll look not only at different kinds of surfaces as a source for electrons but also at the kinds of surfaces that are affected by a charged balloon.

Materials

1   Balloon, 9 inches, round
1   Pile of hair, fine
    Low humidity, preferably
    Variety of other surfaces

Procedure

1. Inflate the balloon and tie it off. Hold the inflated balloon against the chest or back of the person who is donating the pile of hair and see if you can get the balloon to stick there.

2. Hopefully nothing happened. As with the previous experiment, take the balloon and rub it back and forth vigorously on the head of the person with the fine hair. After 15 to 20 rubs, gently lift the balloon off the top of the volunteer's head.

3. Take the balloon and gently place it against a wall. Let go of the balloon and see if it now sticks to this new surface. The first half of this experiment is to find other electron donors. You want to look for items that have a lot of surface area exposed. For example, wool sweaters, flannel blankets, fake fur jackets, and shag carpeting to name a few. Make a list of 10 things and test them by trying to stick the balloon to the wall. Record your results.

Data & Observations

1. Try 10 different surfaces and check whether they are good electron donors or not.

Electron Donor	Yes	No
1. _____	___	___
2. _____	___	___
3. _____	___	___
4. _____	___	___
5. _____	___	___
6. _____	___	___
7. _____	___	___
8. _____	___	___
9. _____	___	___
10. _____	___	___

2. Charge a balloon using hair and then test 10 different surfaces to see if a charged balloon will stick to it or not.

Surface	Yes	No
1. _____	___	___
2. _____	___	___
3. _____	___	___
4. _____	___	___
5. _____	___	___
6. _____	___	___
7. _____	___	___
8. _____	___	___
9. _____	___	___
10. _____	___	___

Stuck-Up Balloons

How Come, Huh?

Same explanation as the previous experiment. When you rub the balloon on the head of the person, the rubber in the balloon attracted the electrons from the hair. This does two things. One, it steals negatively charged electrons from the hair strands; this leaves the hair with a net positive charge; and two, stealing electrons increases the negative charge on the balloon.

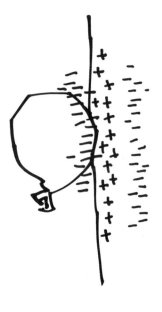

We also know that like charges repel. A negative charge will repel, or push against, other negative charges. When you placed the balloon on the wall or against the body of the person helping you, the electrons in the balloon pushed the electrons in the wall or the body away. This exposed the protons, positively charged centers, of those wall or body atoms. When the balloon with a huge negative charge comes in contact with a wall with positively charged protons exposed, you have an attraction.

Science Fair Extensions

4. Create an inventory of materials that do and do not accept electrons well. Try wood, glass, metal, fabric, plastic, and organic matter like trees and people. Determine if one type of material works better than another.

5. Record the amount of time that a balloon will be stuck to a wall as a function of the humidity in the air.

Sticky Ghost Poop

The Experiment

All right, let's explain the origin of the name of the lab for starters. I was teaching an after-school program to a class full of third graders, and we were doing a chemistry experiment testing the solubility of different kinds of packing peanuts in water and acetone.

During the course of the lab one of my students approached me and said, "You know, Mr. Hixson, this isn't really a Styrofoam packing peanut." Curious, I asked him what he thought it was, and he replied, "Ghost poop," and I fell out of my chair laughing and have been calling it that ever since. Hope you aren't offended.

Back to the lab at hand. We know that electrons can be stolen from lots of different surfaces. We know that electron-rich objects like balloons can stick to lots of different surfaces. The question at hand now is, "How many different objects can be charged and stuck to the wall?"—the neighbor's cat not withstanding.

Materials

1   Bag of Styrofoam peanuts
1   Pile of hair, fine
      Low humidity, preferably

Procedure

1. Take a Styrofoam packing peanut and rub it back and forth vigorously on the head of the person with the fine hair. After 15 to 20 rubs take the packing peanut and place it against a clean, smooth surface. It should stick to the wall.

2. Continue adding packing peanuts to the wall until you spell your name or a greeting of some kind, like, "Science is too cool!"

Sticky Ghost Poop

3. Now comes the fun part. You get to find 10 other items that stick to the wall. You might want to keep in mind that the two things that we have been experimenting with—balloons and Styrofoam packing peanuts—are very light, so you might want to head in that direction.

However, you have also probably rolled in a pile of fresh-out-of-the-dryer clothes and had socks stick to you as well. Don't limit yourself too much. Think large surface area too.

Data & Observations

1. Try 10 different items and check whether they stuck to the wall or fell off.

Item Charged & Tested	Yes	No
1. _____	____	____
2. _____	____	____
3. _____	____	____
4. _____	____	____
5. _____	____	____
6. _____	____	____
7. _____	____	____
8. _____	____	____
9. _____	____	____
10. _____	____	____

How Come, Huh?

As you probably found out, a lot of different materials can be charged. Getting them to stick to the wall is another thing. It turns out that it is a combination of a lightweight object that can also carry a large charge.

The large charge is a function of a large surface area—so anything that has lots of sticky-out parts, like fibers on a sock, works well. That charge then has to be great enough to overcome the pull of gravity on the weight of the object. If it can do that, you have stickage. If the object is too heavy for the charge, it will fall immediately or just plain not stick.

Science Fair Extensions

6. By using a little bit of acetone, available from any hardware or paint store, you can stick Styrofoam packing peanuts together.

Experiment and make a chain or two-dimensional design from the peanuts and see if there is a point where the weight of the peanuts is so great that it overcomes the force of the electrical charge.

7. Go big: Try and stick a flannel sheet to a clean, bare, plasterboard wall.

Pepper Picker-Upper

The Experiment

We are going to ride this horse 'til it drops. We know that electrons can be stolen from a variety of surfaces. We know that a variety of items can be charged and stuck to a number of different surfaces. Now we are going to look at the ability of a balloon to attract and influence a variety of objects.

Materials

1   Pepper shaker
1   Tart pan
1   Balloon, 9 inches, round
1   Pile of hair, fine
    Variety of other objects
    Low humidity, preferably

Procedure

1. Add enough pepper to the tart pan to cover it lightly.

2. Inflate a rubber balloon and tie it off. Rub it back and forth vigorously on the head of the person with the fine hair and steal a big pile of electrons.

3. Bring the charged balloon near but not touching the pepper. Peek under the balloon and observe what happens as you wave the balloon back and forth or move it up and down. Lab time. Find 10 other items that are attracted to the balloon.

Data & Observations

1. Find 10 different, very small items and test to see if they are attracted to a charged balloon.

	Item Tested	Yes	No
1.	_____	___	___
2.	_____	___	___
3.	_____	___	___
4.	_____	___	___
5.	_____	___	___
6.	_____	___	___
7.	_____	___	___
8.	_____	___	___
9.	_____	___	___
10.	_____	___	___

How Come, Huh?

The huge negative charge on the balloon repels the electrons in the pepper, shoving them off into space. This leaves a positively charged piece of pepper that is attracted to a negatively charged balloon.

The trick is to get small, light pieces of stuff that can respond to a large negative charge. Wood shavings, lint from under your bed, spider webs, dried leaves that have been crumpled, Rice Krispies cereal are ideas that we have tried and they work.

Jumping Wood Shavings

The Experiment

In your race to find different items that will be attracted to a balloon you may have emptied the pencil sharpener or snaked a pile of sawdust from your neighbor's shop. But, just in case you didn't get a chance to experiment with those items and take a look at how electrons migrate and saturate objects, now is your chance.

Materials

1   Balloon, 9 inches, round
1   Pile of hair, fine
1   Pile of fine sawdust
    or pencil shavings
1   Tart pan

Procedure

1. Add enough fine sawdust or pencil shavings to the pan to cover it.

2. Inflate a rubber balloon and tie it off. Rub it back and forth vigorously on the head of a person with fine hair.

3. Bring the charged balloon very near the sawdust. Once the balloon has attracted several hundred pieces of sawdust, flip the balloon over so that the sawdust is on the top side of the balloon. Observe very carefully not only what happens to the sawdust but also how it appears when you flip the balloon over.

Data & Observations

1. Draw a picture of what the sawdust looks like when it is on the surface of the balloon.

2. Describe what happens to the sawdust after it has been on the balloon for awhile. _____

_____

_____.

How Come, Huh?

The electrons on the balloon are free to move around. As they do, some of them race up onto the sawdust, saturating it with a negative charge. Negatively charged balloon in the vicinity of a negatively charged piece of sawdust—something is going to be repelled and go flying.

Raptor Attack

The Experiment

This is a fun, silly lab that kids really get a kick out of doing. You are going to take tissue paper then trace and cut out the outlines of the now very famous dinosaur, the raptor. On your command a pack of wild, half-crazed, and definitely skinny raptors are going to attack your balloon.

Materials

1   Sheet of tissue paper
1   Pair of scissors
1   Tracing pencil
1   Balloon, 9 inches, round
1   Pile of hair, fine

Procedure

1. Lay the tracing paper over the outlines of the raptors on the next page and trace as many raptors as you would like to have in your herd. Cut them out.

2. After you are done copying the outlines, cut them out and place them on the table or counter.

3. Inflate a balloon, tie it off, and charge it up on a pile of hair.

4. Slowly lower the balloon toward the herd of raptors and observe when you get close enough for them to attack.

If your raptors are a little slow to attack, try crumpling them up a bit. Everyone is a little edgier when they are crumpled.

How Come, Huh?

The negative charge on the balloon shoves the negative charges in the paper atoms away. This exposes the positively charged centers of the paper atoms that are attracted to the huge negative charge on the balloon. So, the paper appears to jump up and stick—or in this case, attack the balloon.

Science Fair Extensions

8. You can have an attack of anything: piranhas, tigers, mosquitoes—you name it. Have fun making up different designs.

9. Experiment and see if the size of the paper, or the shape of the paper has any effect on its ability to jump off the table and attack.

Wiggly Water Streams

The Experiment

If you could see water molecules, they would look like Mickey Mouse's head: a big round oxygen atom in the center and two little hydrogen atoms tucked up on top. The illustrations on page 43 should give you the general idea.

This kind of configuration is called a bipolar molecule. The reason chemists call it that is because hydrogen atoms have a positive charge and oxygen atoms are negative. Positive on one end and negative on the other, and you have a molecule that behaves like a little magnet. Since there are two poles, they call it a bipolar molecule, which makes perfect sense to us.

Materials

1 Balloon, 9 inches, round
1 Pile of hair, fine
1 Faucet with running water
   Sink

Procedure

1. Inflate a rubber balloon and tie it off. Rub it back and forth vigorously on the head of a person with fine hair. This will build up a huge negative charge.

2. Turn the faucet on and get a very thin, continuous stream of water dribbling out of the faucet. Bring the charged balloon near the water.

Data & Observations

Draw a picture of you, your balloon, and what the thin stream of water did when you brought the balloon near it.

How Come, Huh?

The negative charge on the balloon repelled the negatively charged oxygen atoms and attracted the positively charged hydrogen atoms. Since water is a liquid, it is free to bend, move, and respond to the electric charge.

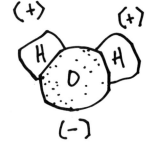

Science Fair Extensions

10. Fill a pie tin full of water and see if bringing a negatively charged balloon near the surface affects the water in any way. See if it is possible to create waves by moving the balloon up and down over the surface of the water.

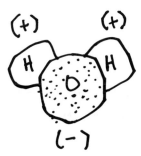

Anti-Gravity Bubbles

The Experiment

You saw from the last experiment that water molecules, arranged like little magnets, are attracted to the large negative charge on a rubber balloon. We are going to extend this one step further.

A bubble is simply a drip of water that has been stretched into a sphere. You are going to defy gravity, amaze the neighbors, and practice good science all at the same time.

Materials

1   Balloon, 9 inches, round
1   Pile of hair, fine
1   Bubble wand with soap
1   Assistant
    Low humidity, preferably

Procedure

1. Inflate a rubber balloon and tie it off. Rub it back and forth vigorously on the head of a person with fine hair.

2. Ask your assistant to remove the cap from the bottle of bubble solution, remove the wand, and blow several bubbles up into the air.

3. Pick one bubble from the herd and hold the balloon just above the bubble. You should notice that the bubble is attracted to the balloon and starts to rise toward the balloon.

4. Your job is to pay attention and move the balloon up and away from the bubbles—the object being to get the bubbles to rise toward the balloon and appear to float in midair.

5. Once you have mastered that technique, then you will deliberately want to attract a bubble, get it to explode on the surface of the balloon, and then bring that same balloon near other bubbles. You should see an entirely different response after the balloon has popped its first bubble.

How Come, Huh?

As we explained in the introduction, a soap bubble is a drop of water that has been stretched out into a sphere. The water molecules themselves are still bipolar, meaning they act like little magnets,

When a large negative charge is brought near the bubble, it attracts the positively charged hydrogen atoms and repels the negatively charged oxygen atom. This attraction to the water molecule is stronger than the downward pull of gravity, so the bubble appears to be rising in defiance of gravity.

Science Fair Extensions

11. If you have access to a Van de Graaff generator, hook yourself up by placing a hand on the sphere. Ask an assistant to blow bubbles toward you and see if enough charge is carried through your body to be able to get bubbles to defy gravity again.

12. Very large balloons, on the order of 10 feet in length or more, are available from toy and novelty stores. Use this huge balloon to attract and collect large bubbles that you have made.

Ping-Pong Ball Obedience School

The Experiment

OK, bubbles are easy. You did it the first time. How about something much heavier? In this lab you are going to use static electricity to attract and move a Ping-Pong ball around a hard, smooth table at your command.

Magic? Nope, just science.

Materials

1 Ping-Pong ball
1 Hard, smooth surface
1 Balloon, 9 inches, round
1 Pile of hair, fine

Procedure

1. Find a nice, hard, smooth, level surface and place a Ping-Pong ball in the middle. If it does not start to roll around, you have selected the perfect place.

2. Inflate a rubber balloon and tie it off. Rub it back and forth vigorously on the head of a person with fine hair. You should be a pro at this by now.

3. Bring the balloon near the Ping-Pong ball but do not touch it. The ball should be attracted to the balloon and start to roll in that direction. Once you get the ball rolling—no pun intended, you can direct it anywhere on the table that you want.

How Come, Huh?

By now, the huge electric charge on the balloon should be extremely familiar to you. This experiment is very similar to one that you did several labs back where you were sticking balloons to the wall. The negative charge on the balloon shoves the negative charges in the Ping-Pong ball atoms away. This exposes the postively charged centers of the Ping-Pong ball atoms—which are attracted to the huge negative charge on the balloon. So, the ball begins to roll . . .

This is where inertia and mechanics take over a bit. Once the ball starts rolling, it has momentum. The attraction to the negative charge simply causes the ball to have more energy and roll faster. As long as the balloon

is tugging at the ball, it will continue to move along a hard, flat surface.

If you remove the balloon, the ball will continue to roll in a straight line until friction and gravity rob all of the momentum from the ball and it eventually stops—until the next charged balloon happens by.

Science Fair Extensions

13. Jumping paper raptors, ghost poop, balloons, and now Ping-Pong balls. We are sure that this is not an exhaustive list. Use your imgination and find other things that are attracted to electric charges and will move, jump, or stick.

Big Idea 3

Static electricity can be observed using a discharge rod, electroscope, piezoelectric rocks, and/or fluorescent lightbulbs.

Mini Lightning Rod

The Experiment

People who live in areas where lightning storms are common often erect lighting rods to direct and ground lightning strikes that come near their houses to prevent damage.

Materials

1   Lump of clay
1   Paper clip
1   Balloon
1   Pile of hair
1   Dark room

Procedure

1. Insert the paper clip in the lump of clay

2. Inflate the balloon, tie it off, and rub it on your head. This will collect a large static charge.

3. Darken the room and bring the balloon near the top of the paper clip. If you have a large enough charge, you will see a spark jump from the balloon to the paper clip.

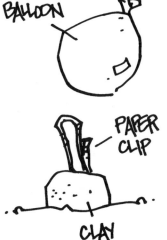

How Come, Huh?

The paper clip is made of metal, which is a very good conductor of electricity. When you brought the static charge on the balloon close to the metal, the electrons recognized a pathway and jumped down to the paper clip.

Lightning rods do the same thing. They provide the path of least resistance for lightning to hit the ground.

Dual Balloon Electroscope

The Experiment

An electroscope is an instrument that detects a weak electrostatic charge. The way that it does this is that the item being tested is touched to the electroscope and passes the charge to the instrument—which has two halves that are flexible and can move.

The next three experiments in this book are variations on this theme with the first one being the easiest and quickest to set up and test. All you need is two balloons, a string, and a pile of very fine hair attached to someone who has a fairly good sense of humor.

Materials

2   Balloons, 9 inches, round
1   Roll of string
1   Ruler
1   Assistant with fine hair, shoulder length

Procedure

1. Inflate both balloons and tie them off. Based on our experience, the fuller they are inflated, the better the chance that the charge will both collect and move around the surface.

2. Cut a 30-inch (75 cm) length of string from the roll. Tie one balloon to each end of the string.

3. Have your assistant hold her finger out at waist level. Hang the center of the string over her finger so that the balloons touch. Observe what happens when they come in contact.

4. Hopefully nothing. There is no charge to repel the balloons from one another. Remove the string and charge both balloons, at the same time, on the head of your assistant. Hang the string over her finger a second time and observe the difference.

How Come, Huh?

The balloons collect a large pile of electrons from the head of your volunteer. The hair passes the charge to the instrument (balloons on a string) that has two halves that are flexible and can move. The charge is dispersed, or spread out, on both halves equally.

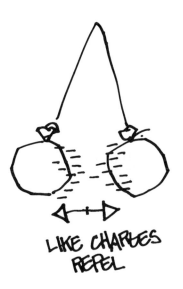

LIKE CHARGES REPEL

At this point the laws of physics take over. If you have two objects with like charges near one another—in this case negative, they repel. So, if the item being tested has a large enough charge, the two halves of the electroscope separate and repel one another. They remain that way until the electrons naturally dissipate or until the top of the electroscope is discharged by the scientist using it.

Science Fair Extensions

14. Change the elements of the electroscope. Use thread instead of string. Try oblong balloons rather than round and see if you get a more dramatic effect.

15. Take a page from earlier in this book and go back and test your electroscope on other surfaces that provide electrons. If this sounds unfamiliar to you head back to page 30 and do the Stuck-Up Balloons lab.

Repulsive Rice Puffs

The Experiment

Balloons are fine but they are very large and relatively difficult to charge. We are going to shift gears and make an electroscope that is easier to charge and one that we can also eat if we get hungry.

Materials

1   Flexible plastic straw
1   Lump of modeling clay
1   Piece of thread, 8 inches
1   Sewing needle
2   Puffed rice kernels
1   Roll of tape
1   Balloon, 9 inches round
1   Volunteer

Procedure

1. Wiggle the lump of clay onto the table and insert the straw. Bend the top part of the straw over to form an upside down "L".

2. Thread the needle through one of the puffed rice kernels and pull the thread through. Tie a knot around that kernel. Thread the needle through the other kernel and tie it to the other end of the thread.

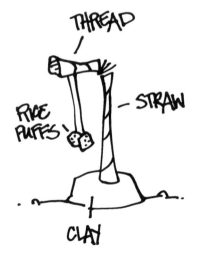

3. Hang the thread on the straw and center it so the rice puffs are touching. Put a small piece of tape on the top of the straw to hold the thread in place.

4. Inflate the balloon, tie it off, and charge it on the head of a volunteer. Bring the balloon in contact with the two rice puffs and transfer the charge to them. Observe what happens.

5. Discharge the rice puffs by touching them. Recharge the balloon and touch it to the top of the straw this time and see if you get the same reaction.

How Come, Huh?

The balloon collects a huge negative charge off the hair shafts. This negative charge is transferred to the rice puffs setting up the same situation that we had in the previous experiment.

Like charges repel. Both the rice puffs had a negative charge so they were repelled by one another. When you touched the puffs, you stole the electrons and they were neutral again.

Finally, when you tried to charge the rice puffs by touching the top of the straw, the experiment showed that the charge was not transferred down the thread to the rice puffs.

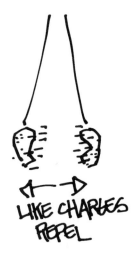

LIKE CHARGES REPEL

Science Fair Extensions

16. Substitute other objects for the rice puffs. Try thin, aluminum foil strips, also called leaves; thin strips of nylon fabric; plastic leaves cut from garbage bags; thin pieces of cellophane; and even strips of rabbit fur.

17. Try charging the rice puffs with different materials—glass rods, fabrics like nylon and wool, and even a Van de Graaff.

Foil Electroscope

The Experiment

This third example is probably the closest to the kinds of electroscopes that early scientists used to study electrostatics and are similar to the commercial brands that are currently available on the market through the dealers that are listed in the front of the book.

Materials

1   Glass jar with lid, 6-10 oz.
1   Nail, small
1   Hammer
1   Paper clip, large
1   Lump of clay
1   Aluminum foil strip, 0.5 inch by 3 inches
1   Aluminum foil square, 4 inches by 4 inches
1   Balloon

Procedure

1. Punch a small hole in the center of the lid with the nail and hammer.

2. Unfold the paper clip and slide it through the hole so that half is above and half is below the hole.

3. Add a lump of clay to the top of the lid to hold the paper clip in place.

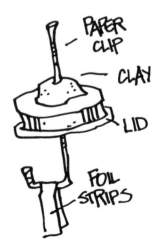

4. Make a small hook on the bottom of the paper clip, the part that will be inside the jar when you put the lid back on. Fold the aluminum strip, 0.5 inch by 3 inches, over the hook. Lower the foil strip down into the jar and screw the lid on.

5. Crinkle the aluminum foil square up into a sphere and gently fix it to the top of the paper clip sticking out of the jar. Your finished electroscope is pictured to the left.

6. Inflate the balloon, tie it off, collect some electrons from your head, and touch the balloon to the metal sphere on top of the electroscope. When an electrical charge is passed from one object to another by direct contact, it is called conduction. You should have noticed the foil strips inside the jar repel, or move away.

7. Take your finger and, watching what happens to the aluminum strips inside the jar, touch the metal ball. The leaves should fall. You are stealing the electrons from the ball and strips and neutralizing the charge. If you are filling out your dictionary as we go, this is another example of conduction.

8. Charge the balloon again, only this time instead of touching it to the metal sphere hold it about a half an inch away and touch the sphere with your finger like the picture to the right. The strips should separate again.

Foil Electroscope

How Come, Huh?

In the first experiment electrons flowed down the paper clip and charged the foil strips, or leaves. Like charges repel so they separated. This is called a charge by conduction, or direct contact from one object to another.

In the second half of the experiment the balloon was nearby but not touching the metal ball. The negative charge in the balloon pushed the electrons in the sphere into your finger, which was acting as an easy pathway, or ground, for the electrons to travel on. By pushing the electrons out of the sphere and paper clip, you left the foil leaves positively charged. Like charges repel so the leaves separated. When an object gets an electric charge without actually touching the source, it is called induction.

Science Fair Extensions

18. Substitute other objects for the aluminum foil strips. Try thin strips of nylon fabric, plastic leaves cut from garbage bags, thin pieces of cellophane, and even strips of rabbit fur.

19. List as many examples of charge by conduction and induction as you possibly can think of. Start with running across a carpet and reaching for a metal door knob.

Piezoelectric Rocks

The Experiment

Many, many moons ago an ancestor of ours took two rocks and whapped them together. When he did, sparks flew and eventually he figured out how to make fire.

The sparks that were created are the focus of our interest. *Piezo* comes from the Greek word that means "pressure." So it stands to reason that a piezoelectric charge is an electric spark produced when two objects bang into or push on one another and create lots of pressure.

Materials

2   Piezoelectric rocks
1   Dark room

Procedure

1. Darken the room.

2. Take two piezoelectric rocks—quartz, tourmaline, and quartzite are good choices—and whap them together fast and hard.

The best way to do this is holding one rock stationary in one hand and hitting it with the other using a quick downward motion striking it with a sideways glance.

How Come, Huh?

In the simplest terms the collision between the two rocks literally scraped the electrons away from the atoms in the rocks and collected them into a charge. When you get a big pile of electrons gathered together, they tend to not like being so crowded so they jump to a larger surface (ground themselves) where they can spread out.

Light It Up

The Experiment

Static electricity can also be used to light up a fluorescent light-bulb that is nowhere near the wall socket or any other source of electricity.

To do this experiment we are going to take advantage of the fact that fluorescent tubes are filled with gases that are easily excited by electricity and electrostatic charges. When they are exposed to electric charges, the electrons in the gases are attracted to the glass that has been coated with a compound called phosphor. When the electrons pass through the phosphor, they produce light.

Materials

1   Milk crate or wooden chair
1   Assistant
1   Dark room
1   Fluorescent lightbulb
1   Rabbit fur

Procedure

1. Stand on the milk crate and ask your assistant to darken the room.

2. Hold the base of the fluorescent bulb with one hand and give the rabbit fur a good rub.

As you rub the fur on the bulb vigorously, look for sparks. Every now and then lift the fur completely away from the bulb and look and listen to what happens.

How Come, Huh?

The reason that you are standing on the milk crate is to insulate you from the Earth, which is a very good conductor. Doing this allows you to build up a larger, more dramatic charge.

By rubbing the rabbit fur on the lightbulb, you are collecting electrons from the glass. This leaves the rabbit fur negatively charged and the surface of the glass bulb positively charged.

The positively charged surface of the glass attracts electrons from the gases trapped inside the bulb to the outside, but to get there they must pass through a coating on the inside of the fluorescent bulb call phosphor. It is the fine, whitish powder that you see if you have ever broken a fluorescent bulb open.

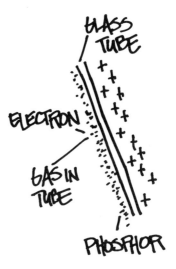

As the electrons pass through the phosphor, they excite it and this produces a visible glow that you can see. This glow can also be produced when electricity jumps from negatively charged rabbit fur back to the positively charged glass bulb. When the discharge hits the bulb, it excites the phosphor, and the bulb glows again.

Science Fair Extensions

20. Zip ahead in the book and find the Van de Graaff section. There is a lab called Static Fluorescence that expands on this idea.

21. Again, there are a variety of materials that will create a static charge: plastic wrap, cellophane, nylon, wool. Experiment and see what works best for you in your area.

Big Idea 4

Static electricity can be collected with an electrophorus and stored in a device called a Leyden jar. The charge can be released from the Leyden jar by contact with any conductive material.

The Electrophorus

The Experiment

One of the more famous of the early scientists who experimented with electricity was an Italian by the name of Alessandro Volta. In 1775 he invented an instrument called an electrophorus— a device that can collect and carry a fairly large electrical charge. The electrons would only leave the electrophorus when they came in contact or near an object, like a human, that would act as a ground and allow the electrons to travel and spread out.

The point of this lab is to build an electrophorus, which you can then use with previous as well as subsequent labs when a static charge is needed.

Materials

1   Candle, short, thick
1   Book of matches
1   Pie tin, aluminum, 9 inches or so
1   Thumbtack
1   Record, old, vinyl
1   Piece of wool, 6 inches by 6 inches
   Adult supervision

Procedure

1. Either with adult assistance or supervision, light the candle and drip wax into the center of the aluminum pie tin.

When there is a puddle of wax, blow out the candle and wiggle the flat, unlit end of the candle into the wax and hold it there while the wax cools and solidifies.

CANDLE

PIE TIN

THUMBTACK

The Electrophorus

2. Insert a thumbtack through the bottom of the pie tin to hold the candle in place.

3. Charge the surface of the old record by rubbing it with the piece of wool. Holding the wax candle, place the aluminum pie tin directly on top of the old record that has been charged.

4. Quickly touch the edge of the pie tin. This grounds the pie tin and allows the negative charge to move from the record up to the pie tin.

5. Remove the pie tin from the record. It is now charged and ready to go. You can use it to light fluorescent lightbulbs or to charge your Leyden jar, which is the next experiment.

Data & Observations

1. Try 6 different items and check whether they can charge your electrophorus or not.

Item Charged & Tested	Yes	No
1. _____	____	____
2. _____	____	____
3. _____	____	____
4. _____	____	____
5. _____	____	____
6. _____	____	____

How Come, Huh?

When the record was rubbed with the wool fabric, it was being given a negative charge. Placing the pie tin on the record meant that the two materials were in contact and conduction was possible, but the electrons on the record have to have somewhere to go and the pie tin appeared full to them.

Touching your finger to the edge of the pie tin allowed the electrons in the pie to jump onto your finger and the electrons on the record to jump onto the pie tin. Because more electrons

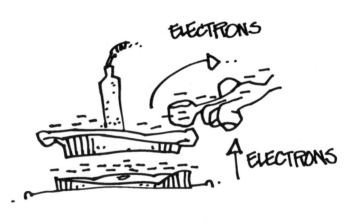

moved from the record to the pie tin than from the pie tin to your finger, the result was a net negative charge on the pie tin—as easy as taking candy from a baby.

Science Fair Extensions

22. Experiment with other sources of electrons. Instead of an old record try a Styrofoam plate, a pile of wool, a television, a radio—or walk across a carpet in wool socks and touch the electrophorus with your finger. Experiment, explore, poke around, have fun.

The Zap Pak

The Experiment

In the previous seven experiments we have been collecting and playing with piles of electrons. They zip all over the surface of the balloon, are transferred to other surfaces, and basically behave like a bag full of marbles that have been emptied out onto the gym floor. The question that will eventually be asked is, "What if we want to save these electrons?" Good question. Simple answer. Make a corral and herd them into it.

Folks who work with electrons as a profession would also call this corral a capacitor. It is an electron storage device that allows you to collect electrons and then use them to make sparks, drive electrostatic motors, or simply get the cat to quit sitting on your lap while you watch television.

One word of caution: There are several recorded incidents where scientists made huge jars and collected huge numbers of electrons. When they released (discharged) the electrons from the jar, the zap that they got was so powerful that it knocked them off their feet and across the room. In fact, Benjamin Franklin almost killed himself using a Leyden jar. Be careful.

Materials

1   Sheet of aluminum foil
1   Pair of scissors
1   Ruler
1   35 mm Film container with snap lid
1   Balloon, 9 inches, round
1   Roll of transparent tape
1   Nail, 6 penny, ungalvanized
1   8-inch Length of 22 gauge copper wire, stranded
1   Piece of flannel fabric, 6 inches by 6 inches
1   Piece of PVC pipe, 2 feet long,
        0.75 inches in diameter
    Water
    Adult Supervision

Procedure

1. Cut two pieces of aluminum foil, 1.5 inches tall by 5 inches long. Roll one piece of the aluminum foil up and place it around the outside of the 35 mm film container.

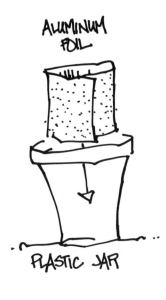

Your capacitor will hold a larger charge and produce a louder, longer spark if you press the aluminum foil against the outside of the container and make it as smooth as possible. Once you get the piece of foil for the inside of the film container as wrinkle-free as possible, tape it in place.

2. Using the tip of the nail, pierce the center of the lid of the container. Push the nail halfway through the hole.

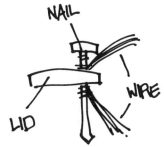

3. Wrap the 8-inch piece of copper wire around the middle of the film canister and aluminum-foil coating.

4. Twist the wire together to hold it in place and make sure that you have enough left over to reach the top of the nail sticking out of the lid. Use the illustration to the right as a guide.

5. Fill the film canister half full with water and snap the lid onto the container.

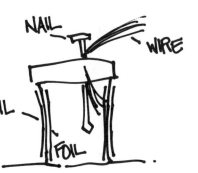

The Zap Pak

WIRE

CAPACITOR

6. Bend one end of the 8-inch copper wire to form a small eye. Use the illustration as a guide. Bend it up toward the head of the nail. You will want to leave a small distance between the "eye" of the wire and the head of the nail. This is called the "spark gap."

7. To charge the capacitor, rub the PVC with the piece of flannel fabric. Use long strokes. As the fabric moves across the plastic tube, it loses electrons to the tube. These electrons are running around looking for someplace to go. To herd them into the capacitor, hold the PVC pipe next to the feathered strands of copper wire leading away from the nail. The electrons will zip down the wires, into the container, and onto the aluminum foil wall inside the capacitor. The illustration below shows an alternative way to charge the Zap Pak.

8. As the number of electrons inside the capacitor increases, things start to get crowded. When there is enough of a charge built up, the electrons discharge, or race up the nail, and jump toward the eye of the copper loop.

How Come, Huh?

A capacitor can be defined as two conductors separated by the plastic wall of the container. When you charged the PVC pipe, the electrons ran down the pipe and into the capacitor.

Repeated rubbings added more and more electrons. These electrons generated a large negative charge that repelled the electrons on the aluminum foil on the outside of the container, causing that layer to have a positive charge. Huge negative charge on the inside and an equally large positive charge on the outside and you have what is called an *electrical potential* built up.

Eventually the electrons on the inside can't stand it anymore and they decide to escape from the capacitor. To do this they have to jump from the head of the nail to the eye, which takes energy. As these electrons are jumping, they lose energy to molecules in the air, which produces light.

When the electrons jump, they heat the air rapidly—just like a mini-lightning bolt—and produce a blue spark. The spark is caused by electrons in the air being excited, which produces light, and the heating and rapid expansion of the air causes a zap, or pop, that can be heard if you listen carefully. Thunder and lightning on command—next thing you know you will be nominated to be a Greek god or goddess for a day.

Science Fair Extensions

23. Experiment with the size of the jars within reason. Benjamin Franklin would be the first to endorse moderation in your efforts to super-size this lab.

The Zap Pak

24. You can also make a static-charge detector out of a 35 mm film canister. Follow the following directions:

A. Punch a hole in the top and bottom of a clear film canister.

B. Flatten the wires from a mini neon bulb. Thread one wire through the bottom and flatten it to keep it in place. Thread the other wire through the top.

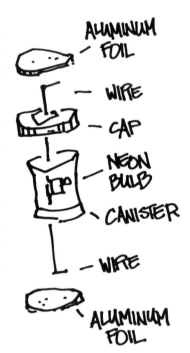

C. Cut two round pieces of aluminum foil and glue them to the top and bottom of the film canister.

D. You have successfully completed a static-charge detector. Walk up to the television and touch one end of the detector to the screen. If the neon bulb lights up, you have detected static electricity.

Big Idea 5

A Van de Graaff generator is a machine invented to produce large amounts of static electricity over a short period of time.

The Van de 'Do

The Experiment
You are going to create the world's wildest hairdo courtesy of several billion electrons on the loose. An assistant, one who preferably has fine, shoulder-length hair, is going to place her hand on the top of the Van de Graaff machine. When you flip the switch, the electrons will start to flow and upon completely saturating your assistant's body will produce a hairdo that stands out among its peers—pun intended.

Materials
1   Van de Graaff generator
1   Discharge wand
1   Milk crate, plastic
    Volunteers, different kinds of hair

Procedure

1. Check to make sure that the machine is off.

2. Ask your assistant to climb up on the milk crate. Make sure that your assistant is not near any electrical conductors like metal legs on chairs or tables.

3. Ask her to place one hand, palm down, on the top of the Van de Graaff.

4. Flip the switch to the Van de Graaff and turn the knob to increase the speed to about 75-80 percent of the maximum. Record your observations. When you have achieved the desired hairdo, have your assistant keep her hand on the top of the generator and turn the machine off. Observe what happens to the hairdo.

5. Ask your assistant to step down off the milk crate anytime after you turn the machine off.

6. Repeat the experiment with kids who have different colors of hair, different thicknesses of hair, and different types of hair—curly, kinky, straight, and wavy.

Data & Observations

Check the box that best describes how much the hair of the volunteers moved when the Van de Graaff generator was turned on and they were attacked by billions of electrons.

Hair Color/Type	None	Some	Lots
Blonde			
Brunette			
Red			
Black			
Straight			
Wavy			
Curly			
Kinky			

The Van de 'Do

How Come, Huh?

What you should have found was that kids with fine, straight hair had the largest reaction to the Van de Graaff generator.

The Van de Graaff generates billions of electrons that are in a very mobile state. The human body is an excellent conduit for moving electricity around, so the electrons that find their way to the surface of the Van de Graaff race onto the assistant's arm and over to her body.

Once the electrons are on her body, they race all over and saturate everything including the hair shafts. Because electrons have a negative charge, the hair is full of negatively charged electrons, and we know that like charges repel, the hair shafts stand on end trying to get away from one another.

When you take your hand off the Van de Graaff, the source of electrons saturating your assistant's body dries up and the electrons on your assistant's body start to jump ship. As more and more electrons leave, the charge in the hair is reduced and everything starts to head toward normal. When your assistant jumps onto the ground, the remaining electrons are conducted down to the Earth, which flattens the hair.

Science Fair Extensions

25. Hey, how about those guys who get hair transplants, plugs, weaves, and other assorted attachments. Herd a bunch of old, formerly bald guys into the lab and see how they do with extra electrons.

Electric Octopi

The Experiment

We saw with the Van de 'Do that hair shafts can be charged to stand on end and fly through the air. What about something more dramatic? This experiment allows you to construct an octopus out of paper and get it in a state of constant motion.

Materials

1   Pair of scissors
1   Section of tissue paper
1   Roll of masking tape
1   Van de Graaff generator
1   Discharge wand
1   Paper bag from grocery store
1   Plastic bag
1   Skein of yarn
1   Spool of string
1   Handful of plastic straws

Procedure

1. Cut 8 long strips of tissue paper 1 inch wide and 24 inches long.

2. With the machine off, tape the ends of all eight pieces of paper to the top of the Van de Graaff generator.

3. Flip the switch to the Van de Graaff and turn the knob to increase the speed to about 50 percent of the maximum.

Electric Octopi

4. Once the octopus is up and floating around the sphere, increase the speed of the Van de Graaff to 80 to 85 percent of maximum and see how that changes the appearance of the paper.

5. Discharge the sphere with the discharge wand and notice what happens to the octopus.

6. Turn the Van de Graaff generator off. Without removing the pieces of paper, cut each of them in half so that the arms of the octopus are thinner than when you started.

7. Turn the Van de Graaff back on and observe the appearance of the octopus with the newly thinned arms. Compare that with the first octopus that you charged.

8. Once you have made your fat and skinny octopi, make other octopi out of grocery bags, plastic bags, yarn, and plastic straws.

9. The last is more like a sea urchin than an octopus. Figure out a way to bundle the straws together and attach them to the top of the dome of the Van de Graaff. If you use light straws, you get a very funny and entertaining result.

10. With all of this taping of things to the dome, be sure to use some rubbing alcohol and give it a good cleaning.

Data & Observations

Check the box that best describes how much the material moved when the Van de Graaff generator was turned on and it was saturated by all those electrons.

Material	None	Some	Lots
Tissue Paper			
Grocery Bag			
Plastic Bag			
Yarn			
String			
Straws			

How Come, Huh?

The electrons race around the sphere and up onto the paper. Each of the eight pieces of paper accumulates a negative charge very quickly.

Like charges repel so the negative charge on one piece of paper is repelled by the sphere—which shoves it into the air. Once it is in the air, it is repelled by the other negatively charged pieces of paper. Because the charges are constantly moving around on the paper, the pieces of paper are subject to different amounts of charge and as a consequence move around a lot. Depending on the weight of the material, you will see the same effect repeated.

Spontaneous Confetti

The Experiment

This is a great activity the next time you have a surprise party and you are a little short on attendees to throw confetti into the air to celebrate the occasion.

Materials

1   Van de Graaff generator
1   Hole punch
1   Pile of colored construction paper scraps
1   Pile of pencil shavings
1   Pile of fabric squares
1   Pair of scissors
1   Pile of packing peanuts
1   Pile of pennies

Procedure

1. Using the hole punch and the colored construction paper scraps, make several hundred colored paper "holes."

2. With the machine off, pile as many of the colored dots on the top of the Van de Graaff generator as you can.

3. Flip the machine on and the confetti explodes into the air. Record your observations in the Data and Observations section.

4. Repeat the experiment using the materials listed above. Cut squares of fabric, crush packing peanuts into little pieces, and pile pennies.

Data & Observations

Check the box that best describes what happened when the pile of material was saturated with electrons.

Material	Exploded	Just Sat There
Paper Holes		
Pencil Shavings		
Fabric Squares		
Packing Peanuts		
Pennies		

How Come, Huh?

The electrons race up onto the dome of the Van de Graaff generator giving it a huge negative charge. The electrons don't stop at the dome, they also saturate the paper holes giving them a huge negative charge. Like charges repel, as we have seen, so the paper holes are repelled off the dome and into the air. The exact same thing happens with everything else but the pennies—they are just too heavy.

If you want to have some fun, ask one of your friends to stand near the generator and try to catch the paper as it flies off the dome. Pile the paper holes on top. Ask them to get ready. Flip the switch on and it's a safe bet that they don't catch a single dot.

Group Shock Therapy

The Experiment

Three (or more) kids are lined up, standing on milk crates, holding hands. When you hit the juice, you are going to turn them into a large human circuit. After you have charged them for a while, you will also demonstrate that a charge can be taken from any point on the circuit without any trouble at all—much to the chagrin of your lab assistants.

Materials

1 Van de Graaff generator
3 Milk crates, plastic
3 Volunteers
1 Discharge wand

Procedure

1. Check to make sure that the machine is off.

2. Ask your assistants to climb up on the milk crates and make sure that they are not near any electrical conductors like metal legs on chairs or tables. They should be in a line. Use the illustration below to help you set this up.

3. Ask the first assistant to place one hand, palm down, on the top of the Van de Graaff. Then have all three assistants join hands forming a chain or open circuit.

4. Flip the switch to the Van de Graaff and turn the knob to increase the speed to about 75-80 percent of the maximum. As soon as the electrons start to flow, they will zip from the person touching the top of the ball all the way down to the end of line and back.

5. Once a charge has been allowed to accumulate, walk to the end of the line and using either the discharge wand or the knuckle of your hand touch the free hand of the last kid in line. You should see and feel a shock.

6. Walk to the middle assistant and take a charge off their elbow and at the same time watch what happens to the hairdos of all three of your volunteers.

7. Using your knuckle or the discharge wand, experiment with taking charges off any and all three of the volunteers. Experiment to see if one place is better than another for collecting charges. Try taking charges from both or all three of the volunteers at the same time.

8. Your assistants can step down off the milk crates anytime after you turn the machine off.

Group Shock Therapy

ELECTRICAL CHARGE BY CONDUCTION

How Come, Huh?

The Van de Graaff generates billions of electrons that are in a very mobile state. The human body is an excellent conduit for moving electricity around so the electrons that find their way to the surface of the Van de Graaff race onto the assistant's arm and over to his body and then onto the bodies of the other two assistants.

Once the electrons are on their bodies, the electrons race all over and saturate everything including the hair shafts. Because electrons have a negative charge, the hair is full of negatively charged electrons, and we know that like charges repel, the hair shafts stand on end trying to get away from one another.

As the electrical charge builds up on the bodies of the three kids, it is also looking for an exit, an off-ramp to a place that is not quite so crowded. So, when another conductor comes along—your knuckle or the discharge wand—the electrons leap, literally, at the chance to get off the over-crowded trio. When they do, the assistants' hairdos drop just a bit with the loss of the electrons, but once the charge builds up again, the hair shafts are up and repelling again.

Science Fair Extensions

26. Repeat the experiment but do not insulate the participants and see how the intensity of the charge is affected.

27. See how long a chain you can make. Be sure to insulate everyone as you add them to the chain. When you get the chain to be as long as you possibly can, have everyone step down, ground themselves, and see how that affects the experiment.

28. Insert a non-conductor, or insulator, into the hands of two of the individuals in the chain and test for the size of the charge on both sides of the insulator. Determine if the insulator does indeed prevent the movement of electrons or if they jump across the barrier anyway.

29. With lots of people, make a chain that has branches. Do the branches weaken, strengthen, or have no effect at all on the strength of the charge that is generated?

30. Make a chain of five people who are all on insulated stands and then ask the middle person to step off the milk crate and ground him or herself. Keep the hands connected, but see how this affects the size of the charge that is produced.

31. Prove that you can take a charge off any person at any place along a chain.

Flying Rabbits

The Experiment

If you are going in order through the book, we have made hair stand on end, created a perpetual octopus, and had a spontaneous celebration complete with a confetti parade. Time to pull out all of the stops and make rabbits fly—or, more specifically, rabbit hides.

Rabbit skins are covered with millions of fine hairs. This gives the rabbit fur a very large surface area to collect things like electrons.

A large surface area means that the rabbit fur can also collect or store a very large negative charge, and we are going to take advantage of that fact to make a rabbit fur appear to fly.

Materials

1   Van de Graaff generator
1   Rabbit fur

Procedure

1. Place the rabbit fur on the dome of the Van de Graaff generator. Center the fur as best you can. The longer it stays on the top of the dome, the more dramatic the "take-off" when it finally flies into space.

2. Flip the machine on and wait and watch as the electrical charge on the fur starts to accumulate.

3. You will notice that the feet, tail, and other thin parts of hide will start to lift off the surface first. When a sufficient charge has accumulated, the rabbit fur will actually lift up off the dome and slide onto the floor—not really flying but we will take it as a good first start.

How Come, Huh?

Again, this is the same idea that we are working with for all of these labs: Two huge negative charges repel one another.

The rabbit fur can collect electrons on each and every one of its hairs—this means that it has room to store bazillions of electrons. The dome generates a huge negative charge; and as soon as the rabbit gets a large enough charge, the two repel one another. The rabbit fur, being the lighter of the two, gets pushed up off the dome. Gravity takes over and pulls the fur to the floor when it gets off center over the dome. If you are really disappointed that rabbits don't fly, we also want to clear up the record about flying squirrels—they only glide, no flying either. It's a cruel world sometimes.

Science Fair Extensions

32. Experiment with other kinds of furs. Once you've made a run through the animal kingdom, try different types of fabric and then finally any kind of material that is altogether flat—paper, plastic, let your imagination run wild.

Static Fluorescence

The Experiment

You have a fluorescent bulb but no socket. No problem. All you need is a source of static electricity. Enter our friend the Van de Graaff generator, a pile of kids, several bazillion electrons—and before you can snap your fingers, you will have light.

Materials

1   Van de Graaff generator
1   Fluorescent tube
3   Milk crates
3   Volunteers

Procedure

1. Check to make sure that the machine is off.

2. Ask your assistants to climb up on the milk crates. Make sure that they are not near any electrical conductors like metal legs on chairs or tables. Arrange your assistants in a line. Use the illustration on page 78 to help you set this up

3. Ask the first assistant to place one hand, palm down, on the top of the Van de Graaff. All three assistants can join hands forming a chain, or open circuit.

4. Flip the switch to the Van de Graaff and turn the knob to increase the speed to about 75-80 percent of the maximum. As soon as the electrons start to flow, they will zip from the person touching the top of the ball all the way down to the end of line and back.

5. Once a charge has been allowed to accumulate, walk to the end of the line and using either end of the fluorescent bulb, touch the free hand of the last kid in line. You should see a spark, and immediately following the spark, the tube should light up.

6. Hand the bulb to the last kid in line and observe what happens to the bulb as the static charge travels through the circuit that you have created.

7. Ask the volunteers to step down off the milk crates and away from the Van de Graaff. Walk up to the machine holding the lightbulb like a saber, step onto one of the milk crates, and carefully bring the bulb near the dome of the Van de Graaff. As the glass nears the dome, the dome should discharge. Observe what happens when the static electricity jumps directly to the bulb.

Static Fluorescence

How Come, Huh?

The inside of the fluorescent bulb is coated with a compound call phosphor. It is the fine, whitish powder that you see if you have ever broken a fluorescent bulb open.

When the glass bulb comes close enough to the dome of the Van de Graaff generator, a portion of the static charge that has built up on the surface jumps to the glass and excites the gases inside the bulb. As the electrons from the excited gas leave the bulb, they pass through the phosphor compound that coats the inside of the glass bulb exciting the electrons in that compound. When these electrons get excited, they produce a bright flash of light that your eyes see.

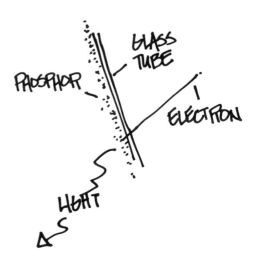

Science Fair Extensions

33. Design an experiment to determine if it matters whether the bulb is "dead" or not.

34. Experiment with other kinds of bulbs and see if it is all bulbs that light up like this or just fluorescent bulbs.

35. Collect a static charge in your Leyden jar using an electrophorus that you made and see if there are enough electrons in either of those two instruments to get the bulb to light.

Big Idea 6

Scientists use symbols to abbreviate the different components of an electrical circuit. These symbols are used when drawing electrical designs called schematics.

Electrical Hieroglyphs

The Experiment

Scientists like to cut corners like everyone else. In the case of drawing electrical circuits they use symbols to represent wires, switches, motors, bulbs, batteries, and other parts of the electrical circuit. This saves time and space. Can you imagine having to draw a motor or battery every time you used it in an electrical circuit? What a pain in the tuchus. This lab will introduce you to some of the basic symbols that are used by electricians.

Materials

1   Pencil

Procedure

1.   In the space to the right are the symbols for wires, batteries, and lightbulbs, used by electricians, electrical engineers, and scientists to design or write electrical circuits.

Data & Observations

1. In the space below are the symbols that describe how to assemble a simple circuit to light a lightbulb with a wire and battery.

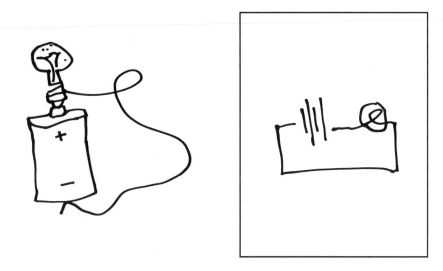

2. Look at the circuits that have been drawn out by hand in the spaces below. Use those illustrations to draw the circuits using electricians' shorthand in the boxes to the right and on page 90.

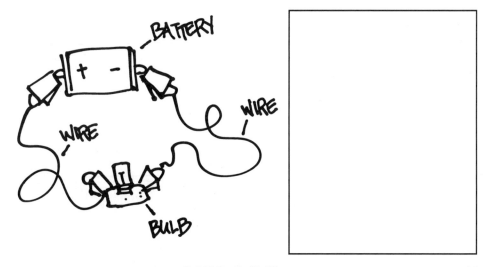

Electrical Heiroglyphs

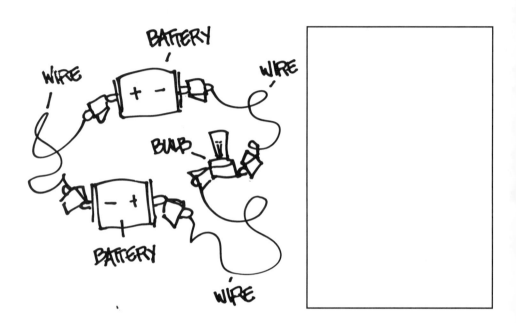

More Symbols

Pictured below are the rest of the symbols that we will use for activities in the remainder of this book.

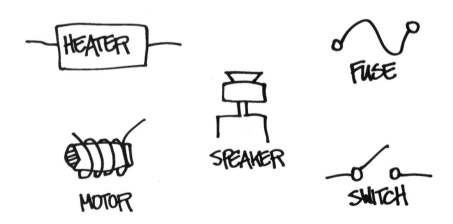

Big Idea 7

Electrical circuits require a complete loop through which an electric current can pass. Circuits can either be wired in series or parallel depending on the use.

A Simple Circuit

The Experiment

Enough of these electrons flying everywhere. We are going to move onto another source of electrons—batteries—and ask them to help us in our pursuit of understanding how electricity works.

To really understand how electricity moves from one place to another you have to develop the very basic ideas, which is the purpose of this lab.

Materials

1   D Battery
1   8-inch Length of copper wire
1   Bulb

Procedure

1. Using one battery, one piece of copper wire, and a single lightbulb, see if you can figure out how to connect all three so that electricity is flowing through the bulb and producing light.

2. On the page to the right are six boxes. In those boxes we would like you to draw at least six different attempts to get the lightbulb lit. Draw each combination, and then build it with your materials to see if the lightbulb can be lit. If the bulb lights, do nothing else to your drawing. If the bulb does not light, put a large "X" over your illustration.

A Simple Circuit

3. If you have tried six times and you still cannot figure out how to light the bulb, that's OK; scientists run into this kind of problem all of the time. You now know six ways that will not work—and that is valuable information in and of itself.

4. Use the illustration to the right to show you one way to get the bulb to light.

How Come, Huh?

Electrons only flow if a circuit is complete. Another way of saying this is: Every piece or part of the circuit is connected in at least two places to the other parts of the circuit.

Another thing that is important to know is that all electrons flow out from the positive (+) end of the battery and back into the negative (-) end. In this experiment that knowledge is not so important, but as you progress through this book, some of the things that you are going to use only work if they are hooked up properly.

Finally, electrons only flow through materials called conductors. These are usually items that are made from metal. If at any point along the circuit, the circuit is interrupted the electricity will not flow.

Science Fair Extensions

36. Prove that the lightbulb will light up either way when you hook it to the battery.

37. Look around your house or school and find other materials that are conductors, like the copper wire. Substitute them for the wire and see if the bulb still lights.

The Domino Circuit

The Experiment

You now know that electricity only flows in a closed circuit. This lab explores one of the two ways that electricity can move through a circuit. This particular pathway is called a series circuit, and it is created when you line each item in your circuit up along a single pathway or electron road. The electrons flow through every item and the electricity is shared by each item.

Using this kind of circuit has the advantage of being very simple to build, but that convenience is outweighed by the fact that it can be easily disrupted and the electricity is shared by all of the pieces in the circuit, so there is less juice for an individual item—all things we will explore.

Materials

1	Switch
5	Alligator clips
1	Battery with battery clip
3	Bulbs with sockets

Procedure

1. Using the diagram to the right, build the circuit that is represented by the symbols. Close the switch and mentally record the brightness of the lightbulb.

2. Add a second lightbulb and alligator clip to your circuit using the diagram on the next page as a guide. When they are in place, close the switch and mentally record the brightness of the lightbulbs. Record your observations in the Data & Observations section.

The Domino Circuit

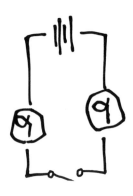

3. Add a third lightbulb and alligator clip to your circuit using the diagram below as a guide. When they are in place, close the switch and compare the brightness of the lightbulbs with the previous two experiments. Record your observations in the data section.

4. Close the switch and unscrew the first lightbulb in the series. Record what happens to the electricity flowing to the other two lightbulbs in the Data & Observations section.

5. Screw the first lightbulb back in and unscrew the second lightbulb in the series. Record what happens to the electricity flowing to the other two lightbulbs in the Data & Observations section.

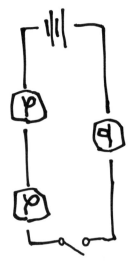

6. Repeat that procedure with the third lightbulb. Record your observations in the Data & Observations section.

Data & Observations

Circle the answer that best describes your observations:

1. When the second lightbulb was added to the circuit, the amount of light given off by both bulbs was _____ compared with just one bulb.

A. dimmer   B. the same   C. brighter

2.   When the third lightbulb was added to the circuit, the amount of light given off by all three bulbs was _____ compared with two bulbs.

        A. dimmer   B. the same   C. brighter

3. When the first lightbulb in the circuit was unscrewed, what happened to the other two bulbs?

        A. They remained lit.   B. They turned dark.

4.   When the second lightbulb in the circuit was unscrewed, what happened to the other two bulbs?

        A. They remained lit.   B. They turned dark.

5. When the third lightbulb in the circuit was unscrewed, what happened to the other two bulbs?

        A. They remained lit.   B. They turned dark.

How Come, Huh?

In a series circuit the electricity flows through every single item in the circuit. The more items in the circuit, the more sharing that goes on, and the less electricity there is for each component. That is why the bulbs got dimmer and dimmer as you added more pieces to the circuit. The other downside to a series circuit is that since the electricity flows through every piece of the circuit, if one of those pieces is faulty or becomes disconnected in any way, the flow of electrons stops dead and nothing works. Until the last couple of years, Christmas trees lights were on a series circuit. If one bulb went on the fritz, the whole strand of lights went out, and you had to test each light until you found the culprit.

Science Fair Extensions

38. Nab an old set of Christmas tree lights and experiment with them. Determine if they are wired in series or not.

Parallel Pathways

The Experiment

The domino or series circuit gets it name because the pieces of the circuit are all lined up like dominos—when one piece falls, or is not able to conduct electricity, the rest of the pieces fall in a line behind the first.

A parallel circuit is different than a series circuit in that it provides the electrons an alternative pathway if the first path gets blocked. This is the kind of circuit that you find in most modern houses. The electricity is in the wires and ready to flow, all you have to do is flip the switch and the light comes on, the stereo fires up, the dishwasher starts to cycle, or the blender starts to spin. With a parallel circuit if something goes down, fails to conduct electricity, or blows up, the rest of the pieces of the circuit keep working—unlike a series circuit where everything comes to a screeching halt.

Materials

6   Alligator clips
1   Battery with battery clip
1   Switch
3   Bulbs with sockets

Procedure

1. Using the electrical diagram to the right, build the circuit that is represented by the symbols. Close the switch and mentally record the brightness of the lightbulb.

2. Add a second lightbulb and two alligator clips to your circuit using the diagram to the right as a guide. When they are in place, close the switch and mentally record the brightness of the lightbulb. Record your observations in the Data & Observations section.

3. Add a third lightbulb and two more alligator clips to your circuit using the bottom diagram as a guide. When they are in place, close the switch and compare the brightness of the light-bulbs with the previous two experiments. Record your observations in the Data & Observations section.

4. Close the switch and unscrew the first lightbulb in the series. Record what happens to the electricity flowing to the other two lightbulbs in the Data & Observations section.

5. Screw the first lightbulb back in and unscrew the second lightbulb in the series. Record what happens to the electricity flowing to the other two light-bulbs in the Data & Observations section.

6. Repeat that procedure with the third lightbulb. Record your observations in the data section. Then pick any two bulbs, unscrew them, and observe.

Parallel Pathways

Data & Observations

Circle the answer that best describes your observations:

1. When the second lightbulb was added to the circuit, the amount of light given off in both bulbs was _____ compared with just one bulb.

        A. dimmer   B. the same   C. brighter

2. When the third lightbulb was added to the circuit, the amount of light given off by all three bulbs was_____ compared with two bulbs.

        A. dimmer   B. the same   C. brighter

3. When the first lightbulb in the circuit was unscrewed, what happened to the other two bulbs?

        A. They remained lit.   B. They turned dark.

4. When the second lightbulb in the circuit was unscrewed, what happened to the other two bulbs?

        A. They remained lit.   B. They turned dark.

5. When the third lightbulb in the circuit was unscrewed, what happened to the other two bulbs?

        A. They remained lit.   B. They turned dark.

6. When any two lightbulbs in the circuit were unscrewed, what happened to the other bulb?

        A. It remained lit.   B. It turned dark.

How Come, Huh?

Since there are multiple pathways, the amount of electricity being drawn from the battery increases when you add a second and third bulb to the circuit. Because of this you did not see much of a change in the amount of light that was produced by each of the bulbs in the circuit as they were added.

Also, in a parallel circuit the electricity has several choices of where to go. If the bridge is out, so to speak, there are alternative paths that can be taken. When you unscrew a single lightbulb, there are other pathways for the electricity to flow, so the flow remains uninterrupted—as opposed to a parallel circuit where everything grinds to a halt. In fact, you can unscrew two or more components, and as long as there is at least one pathway for electrons to flow, they will find it and light things up.

Science Fair Extensions

39. Nab an old set of Christmas tree lights and experiment with them. Determine if they are wired in parallel or series. See if you can figure out how to change them to the other kind of circuit.

40. Exchange bulbs for batteries and see what effect it has on the circuit when you have multiple batteries in series and parallel.

Big Idea 8

Materials that allow the free movement of electrons are called conductors. Those that do not are called non-conductors or insulators.

Electron Herding 101 • B. K. Hixson

Conductivity Tester

The Experiment

Electrons are found in every atom. They are knocked loose and are bumped around all the time, but they do not have the ability to pass through all materials. In fact, they don't travel very well at all in some materials—these are called insulators. The materials that do allow electrons to zip on through are called conductors. This lab will introduce you to a technique that will allow you to test any material and determine if it is a conductor or an insulator.

Materials

1   Battery with battery holder
1   Lamp with socket
3   Alligator clips
1   Nail, 8 penny
1   Copper wire, 18 gauge, 3 inches
1   Craft stick
1   Strip of fabric, 1 inch by 4 inches
1   Glass microscope slide
1   Plastic straw
1   Paper clip

Procedure

1. Snap the battery into the battery holder. Bend the tabs on the ends of the clip out so the alligator clips can be attached easily.

2. Assemble the circuit pictured to the right and make sure that the bulb lights up when everything is connected together.

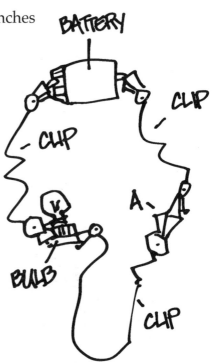

Conductivity Tester

3. Separate the two alligator clips (point A in the circuit) and test the nail by clipping it between the two alligator clips that you just separated. If the light goes on, that particular item is a conductor; if it remains unlit, it is an insulator.

4. Test each of the items in the data table one at a time by clipping them into the circuit. Mark the appropriate box in the data table below.

Data & Observations

In the data table below identify the material—wood, glass, plastic, metal, and so on—for each of the items below and then also mark whether each is an insulator or a conductor.

Item	Material	Conductor	Insulator
Nail			
Copper Wire			
Craft Stick			
Fabric			
Microscope Slide			
Plastic Straw			
Paper Clip			

How Come, Huh?

What you should have discovered is that metal objects make excellent conductors, while wood, glass, plastic, fabric, and other kinds of materials do not.

The main difference between these two is that the metals have a very orderly, simple, crystalline structure; whereas the others have more complex arrangements of molecules. The simplicity of the design for metals allows the electrons to flow freely through them, passing from atom to atom.

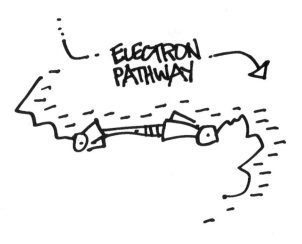

Science Fair Extensions

41. Expand the list and test more items from around your home. Include natural fibers like hair, leaves, fur, and things that were left off the original list.

42. Design and build a portable conductivity tester.

Liquid Conductors

The Experiment

OK, we've got the solid conductors covered. We know that sparks will jump through the air, so we'll call it good for the gases too. What about liquids? You always hear that you should not use electrical devices when you are in the tub, or don't shower and shave with an electrical shaver plugged into the wall. This lab does two things: explores conductivity of liquids and sets you up to understand electroplating.

Materials

1   Battery with battery holder
1   Lamp with socket
3   Alligator clips
1   250-ml Beaker
    Water
    Salt, 2 tablespoons
    150 ml Lemon juice
    150 ml Vinegar

Procedure

1. Snap the battery into the battery holder. Bend the tabs on the ends of the clip out so the alligator clips can be attached easily.

2. Assemble the circuit pictured to the right and make sure that the bulb lights up when everything is connected together.

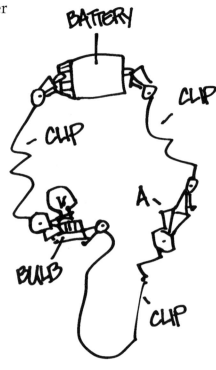

3. Separate the two alligator clips (point A in the circuit) and test the conductivity of water by lowering the two alligator clips into the water. If the lightbulb lights, the electrons are flowing; if there is not light, it is a poor conductor.

4. Test each of the other three liquids in the data table one at a time by dipping the alligator clips into the beaker. Mark the appropriate box in the data table below. When you get done with our suggestions, add two of your own and test them as well.

Data & Observations

In the data table below mark whether the liquids are insulators or conductors.

Item	Conductor	Insulator
Water		
Salt Water		
Lemon Juice		
Vinegar		
A.		
B.		

Liquid Conductors

How Come, Huh?

What you should have found is that water is actually not a very good conductor. However, if a liquid has an acid in it, such as vinegar (acetic acid), lemon juice (citric acid), or if it dissolves into solution and produces ions (charged particles) like salt, it is a good conductor.

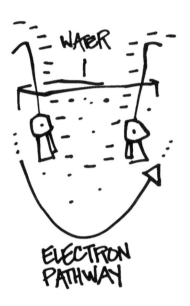

Water is a bipolar molecule, meaning that it has a positive charge on one end and a negative charge on the other, but it not charged itself. It is neutral, or balanced. These other liquids have electrical charges, and this allows them to collect and pass electrons along through them with ease.

Science Fair Extensions

43. With adult supervision or permission collect several acids and bases. Test each of them with your conductivity tester and see if they work equally well—or if one is better than the other.

44. Find out what kinds of chemicals you can add to water to enhance the conductivity and promote the movement of ions and other charged particles through liquids.

Painting with Electrons

The Experiment

Scientists have learned to take advantage of the fact that electricity can flow through certain liquids, and they use that idea to coat objects with metal. The process is called electroplating and is used extensively in industry for all kinds of applications—from making jewelry, to coating machine parts to prevent corrosion, to applying thin magnetic coats to speakers to make them perform more efficiently. In this lab we are going to deposit copper on a paper clip.

Materials

1   D Battery and holder
1   Lamp with socket
3   Alligator clips
1   250-ml Beaker
1   Strip of fabric, 1 inch by 4 inches
    Water
1   Tablespoon
1   1 oz. Bottle of copper sulfate
1   Paper clip
1   Nail, 8 penny
1   Craft stick
1   Glass microscope slide
1   Plastic straw

Procedure

1. Snap the battery into the battery holder. Bend the tabs on the ends of the clip out so the alligator clips can be attached easily.

2. Assemble the circuit pictured to the right and make sure that the bulb lights up when everything is connected together.

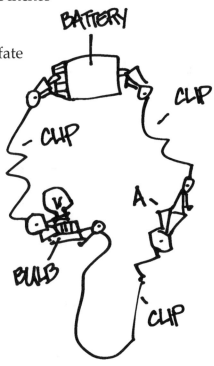

Painting with Electrons

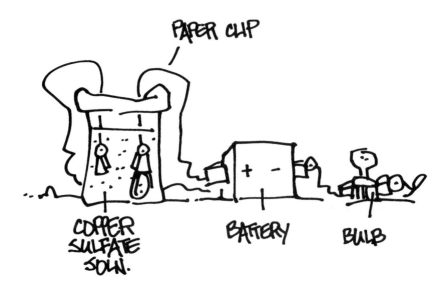

PAPER CLIP

COPPER SULFATE SOLN.

BATTERY

BULB

3. Fill the 250-ml beaker with water and add a tablespoon of copper sulfate. Stir. If it all dissolves, keep adding more until there is a small quantity at the bottom of the beaker.

4. Separate the two alligator clips (point A in the circuit) and clip the paper clip onto the negative lead. Check the battery to see which lead is negative. Lower the paper clip on one side of the beaker and the other alligator clip on the other side. Wait about 3 minutes.

5. When the time is up, pull the paper clip up out of the solution and take a peek at the deposit that is coating the wire.

6. Look at the list of materials in the data table and make a prediction—yes or no—about whether or not the item will take a copper coat. Write your prediction in the box provided. Test all of the other materials that are on the list and see which ones will take a coat and which ones will not. Record your observations in the data table.

Data & Observations

In the data table below identify the material—wood, glass, plastic, metal, and so on—for each of the items below and then also mark whether a coating was deposited or not on that item.

Item	Prediction	Coating? Y/N
Nail		
Craft Stick		
Fabric		
Microscope Slide		
Straw		
Plastic Straw		

How Come, Huh?

What you should have discovered is that metal objects make excellent conductors and are the only materials that were coated by the copper ions.

When you dumped the copper sulfate, it dissolved and broke into charged particles called ions. The copper ions had a positive charge and the sulfate ions had a negative charge. When the two alligator clips were dipped into the solution, one had a positive charge and the other had a negative charge. The negative charge, holding the paper clip, attracted the positively charged copper ions that produced that brownish, gooey coating that you saw on the paper clip.

Science Fair Extensions

45. Find out what other kinds of metals will dissolve in solution and how they can be applied to different types of metals.

Big Idea 9

Electrical circuits can produce light, heat, sound, motion, and magnetic effects. The movement in an electrical circuit is regulated by a switch.

Switches

The Experiment

We've arrived at the point where we know that electrical switches only work if they are connected (or closed). But you can't construct an electrical circuit and have electrons flowing through it willy-nilly all the time. If that were the case, your lights would always be on, the stereo would be blaring, the dishwasher washing, and the garbage disposal grinding. What a racket! What a waste of electricity.

So, to resolve this problem, scientists quickly invented a tool, called a switch, that allows you to control when the electricity is on or when it is off. You use switches all the time. You use them when you turn on the lights in your abode, as you start or stop machines, and when you open or close car doors. You control the flow of electricity.

This lab will reintroduce you to the symbol for a switch. You will also build one from scratch and use it in a simple circuit.

Materials

1   Piece of wood, 2 inches by 1 inch by 3 inches
2   Thumbtacks, metal
1   Paper clip, large
3   Alligator clips
1   Battery with battery clip
1   Bulb with socket

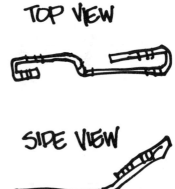

TOP VIEW

SIDE VIEW

Procedure

1. Open the paper clip so that it looks like the illustration to the right. You are going to want to bend the paper clip to an angle of about 20 degrees. That would be the side-view illustration you see to the right.

Switches

TOP VIEW

SIDE VIEW

2. Place the paper clip on the piece of wood and, using one of the metal thumbtacks, secure the narrow half to the paper clip.

3. To complete your switch, position the other thumbtack under the larger half of the paper clip. Push the thumbtack a little more than halfway into the wood. When you push down on this half of the paper clip, it should touch the top of the second thumbtack and then spring back up when you release the pressure. Use the two illustrations to the left as guides.

4. Using the illustration below as a guide, attach one alligator clip to the edge of the paper clip and the other alligator clip to the edge of the thumbtack that is pushed into the wood.

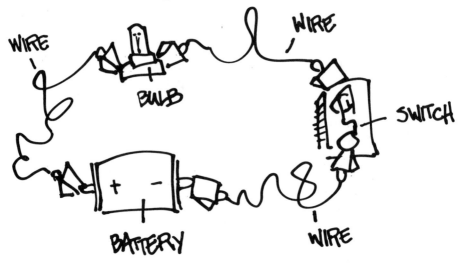

WIRE
WIRE
BULB
SWITCH
BATTERY
WIRE

5. Add a lightbulb, battery, and one more alligator clip to your circuit. When the paper clip touches the second thumbtack, the circuit is completed and electrons can flow freely. Proof of this fact would be the lightbulb coming on. Tap the paper clip on the thumbtack several times so that the light blinks on and off.

Data & Observations

You know what the symbols for lights, batteries, and wires are from the previous labs. The symbol for a switch is pictured in the thought balloon over the cartoon character to the right. Using that symbol, draw your completed circuit in the box below.

Nichrome Wire Lamp

The Experiment

Electricity can be converted to produce sound (very annoying sounds), magnetism, and motion in the form of a motor shaft spinning and a propellor spinning around. Electricity can also be used to generate light, which you saw in one of the first experiments, and heat.

To generate heat from electricity you can hook a nichrome wire up to a battery, 1.5 volts in all, and watch and feel as the electrons zip through the wire, bouncing off atoms as they go, producing heat. It's as simple as that.

Materials

2   Alligator clips
1   Nichrome wire
1   D Battery, 1.5 volt
1   Pencil

Procedure

1.   Using the pencil for a base, leave a one-inch tail and wrap the wire around the pencil several times. Continue to wrap until you have one inch left and stop. Slide the wire off the pencil and gently pull the coil apart so it is a little more loose than when you originally wrapped it.

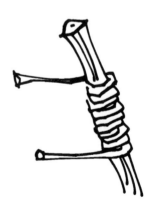

2. The symbol for a heater is pictured below as well as on other pages. Using the schematic drawing to the left below, construct the circuit. When it comes time to hook the heater up, simply attach each end of the coil to an alligator clip and you are good to go.

3. Be very careful when you touch the wire, it will appear warm, and if you use a bigger battery, it will get very hot. Draw the circuit that you built in the box below.

How Come, Huh?

Nichrome wire is the same wire that is used in heat dishes, toasters, toaster ovens, and other appliances that convert electrical energy to heat. The molecular structure of the wire is such that as the electrons bounce through the wire they get excited and produce not only heat but also light. Two things that we see and feel.

Electron Convection

The Experiment

This is a great demonstration to actually show the movement of the cold and warm currents in water. Warm water will rise when heated in the aquarium, which in turn will bend light. There right before your eyes, waving about on the big screen, you will be able to see convection in motion through your convection window.

Materials

1   Small aquarium
    Water
1   Nichrome wire, 6 inches
1   Pencil
2   Alligator clips
1   6-volt Battery
1   Flashlight
1   Bottle of food coloring
1   White, wall surface

Procedure

1. Fill the aquarium full of water. For this experiment it is better that the water be cool to cold so that the convection currents are more dramatic.

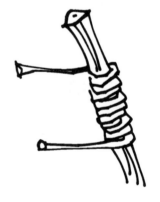

2. Take the nichrome wire and wrap it around the pencil, leaving a one-inch tail on each end. Remove the pencil when you are done wrapping.

3. Put an alligator clip onto both ends of the nichrome wire. Lower the nichrome loop (horizontally) into the center of the tank full of water. Connect the other end of the leads to the battery.

4. Place the flashlight in front of the screen and prop it up so that it shines just above the nichrome loop in the tank and onto the back wall.

5. As the electricity starts to flow through the wire, it will become hot. The hot wire will immediately pass the heat, through conduction, to the water molecules that will begin to get warm. The warm water will be pushed up by the heavier, colder water and start to rise in a convection current. Since the water is less dense, the light passing through the warm water will be bent to a different angle than the light passing through the cold water. This produces the effect of waves and shadows that you can see on the wall behind the aquarium.

6. Add a few drops of food coloring to the water and the convection current becomes even more visible. Draw what you see in the box on the next page.

Electron Convection

Data & Observations

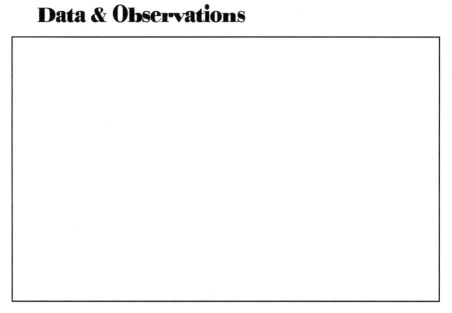

How Come, Huh?

Hot air rises! You know that because you have probably sat at the top of the bleachers during a basketball game before and started to sweat. The people closer to the floor seemed quite comfortable. The same happens with water. Water will expand as it gets warmer and it will become less dense. The warm water will rise up above the cooler water, and this will set up a convection current. The screen allows you to see the bending of the light waves as they pass through the warm water. Adding food coloring to the aquarium allows you to actually see the warm water rising to the top.

Science Fair Extensions

46. Experiment and see if the voltage of the battery affects the speed that the convection current rises to the top of the aquarium.

47. Try other conductors in place of the nichrome wire loop: For example, nails, wire, copper, and a lightbulb come to mind. Think outside the current.

What's the Buzz?

The Experiment

We are going to continue producing sound from a variety of sources of energy using electicity and a simple buzzer to accomplish the task. The trick to this buzzer is that it only works when electricity is flowing through it in the proper direction, which is the first thing that you are going to figure out. Then once you get the annoying buzz that is characteristic of this instrument, we will introduce you to the Morse code.

Materials

1   Switch
3   Alligator clips
1   Buzzer
1   Battery with battery clip
1   Friend

SPEAKER

Procedure

1. Using a battery, 2 alligator clips, and the buzzer, build the simple circuit that is pictured to the left. As you connect the buzzer to the battery, one of two things will happen. You will either hear an annoying buzz or you will be greeted with silence. If you hear the buzz, hooray. Circle below which wire, red or black, is connected to the negative side of the battery.

A. Red
B. Black

What's the Buzz?

2. Just to make sure this is a one-way buzzer, reverse the wires and see if you can produce any sound.

3. Finally, build the second circuit, pictured to the left, and use your switch to buzz your friend in Morse code—which is listed on the right-hand side of the page.

4. Using the Morse code, created by Samuel Morse, write a simple message to your friend and then use your buzzer to send it. Ask your friend to write down the letters they think you are trying to send; then compare the message you sent with the message they received. If you have time, switch roles and do it again.

How Come, Huh?
The electricity causes a thin membrane, just like in a kazoo, to vibrate back and forth. The membrane vibrates or pushes against the air causing sound to be produced.

A	. -
B	- . . .
C	- . - .
D	- . .
E	.
F	. . - .
G	- - .
H	. . . .
I	. .
J	. - - -
K	- . -
L	. - . .
M	- -
N	- .
O	- - -
P	. - - .
Q	- - . -
R	. - .
S	. . .
T	-
U	. . -
V	. . . -
W	. - -
X	- . . -
Y	- . - -
Z	- - . .
0	- - - - -
1	. - - - -
2	. . - - -
3	. . . - -
4	. . . . -
5	. . . . .
6	- . . . .
7	- - . . .
8	- - - . .
9	- - - - .

Message you wish to send:_____

_____.

Letter	Morse Code for Letter	Morse Code	Letter
_____	_____	_____	_____
_____	_____	_____	_____
_____	_____	_____	_____
_____	_____	_____	_____
_____	_____	_____	_____
_____	_____	_____	_____
_____	_____	_____	_____
_____	_____	_____	_____
_____	_____	_____	_____
_____	_____	_____	_____
_____	_____	_____	_____
_____	_____	_____	_____
_____	_____	_____	_____
_____	_____	_____	_____
_____	_____	_____	_____
_____	_____	_____	_____
_____	_____	_____	_____

Science Fair Extensions

48. Research decibels, deafness, and any correlation between the excesses of one and the result of the other.

49. With your parents' permission, take apart a speaker and try to figure out its parts and what makes it tick.

Nail Magnets

The Experiment

Take an ordinary nail, wrap it with several loops of copper wire and hook a battery to it, and you have an electromagnet—a magnet that was created when electricity flowed through a wire around an iron core.

Electromagnets and the other ideas that you are learning are the foundation for all modern-day electric motors. Cars start and run because they have an electromagnet in the motor. Most countertop kitchen appliances, home repair tools, and any other electrical device that runs with a motor are powered with an electromagnet.

Materials

1 Coil of bell wire
1 Pair of wire strippers/cutters
1 16-penny Nail, ungalvanized
1 Box of paper clips, small
1 Battery holder
1 D Battery
2 Alligator clips

Procedure

1. Cut a 24-inch length of bell wire from the coil and strip both ends. To strip the end of a wire you place about one-half inch of wire in the wire stripper, clamp down and pull the wire through the little hole. The wire stripper should have cut through the plastic but not the metal, and when you pulled, the plastic insulation should have come off, exposing the copper wire inside.

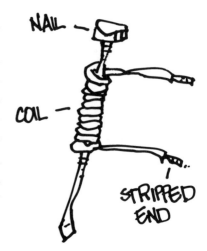

NAIL —

COIL —

STRIPPED END

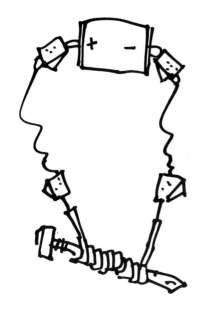

2. Leave a 4-inch tail and wrap one coil of wire around the nail as tight as you can. Add a second coil right next to the first. Do this until you have a total of 10 coils on the nail. Use the illustration on page 124 as a guide.

Lower the nail into a pile of paper clips and see how many you can attract. Hopefully none, but in the spirit of establishing a control we had to try.

3. Snap the D battery into the clip and attach an alligator clip to each end of the battery clip.

4. Attach the loose end of each alligator clip to the bell wire that you stripped to complete the circuit. As soon as the circuit is completed, the electricity will start to flow through the wires wrapped around the nail—and you will have created an electromagnet.

Nail Magnets

5. Dip the electromagnet into a pile of paper clips and see how many you can pick up with your newly constructed electromagnet. Count them and record this number in the data table below in the square that correlates to 10 wraps.

6. Continue to experiment by increasing the number of wraps around the nail in increments of 10. Record the number of paper clips that you pick up each time.

Data & Observations

# Wraps	10	20	30	40	50
# Paper Clips					

How Come, Huh?

We know that when electricity flows through a wire it produces a magnetic field around the wire. This is because electricity is made up of electrons that are zipping through the wire, like water moving through a hose. As the electrons flow through the wire, they are lined up in an orderly fashion producing a magnetic field that organizes the iron particles in the nail. As the iron particles get organized, they produce a magnetic field of their own, which is what attracts the paper clups.

Science Fair Extensions

50. Experiment with different kinds of cores. The question being, "Are all electromagnets built with nails or can we use other objects?" Try a plastic straw, wooden craft stick, steel bolt, aluminum can, or other ideas.

Beannie Toppers

The Experiment

The symbol for a motor is pictured to the right. We are going to move from sound and a simple buzzer through electromagnets, to the things that electromagnets help to produce—motors. This motor is very small and very simple. We are going to not only harness the power in the battery and convert it to mechanical energy, but we are also going to attach a propeller to the shaft

MOTOR

of the motor and provide you with the rare opportunity to construct your own self-propelled beanie. It won't get any better than this until you hit junior high.

Materials

1   Homemade switch
3   Alligator clips
1   Motor
1   Plastic propeller
1   Battery with battery clip

Procedure

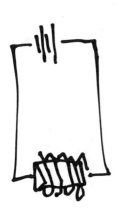

1. Using a battery, 2 alligator clips, the motor and the propeller, build the simple circuit that is pictured to the left. As you connect the motor to the battery, one of two things will happen. The propeller will begin spinning either clockwise or counterclockwise. Check the wires and fill in the data table on the next page.

2. Reverse the wires and fill in the rest of the data table. You now have proof positive that electricity flows through wires in one direction and that electrical energy can be converted into mechanical energy.

Beannie Toppers

Data & Observations

Motor Spins	Lead Connected to (+) Terminal
Clockwise	
Counterclockwise	

How Come, Huh?

When electricity flows through a wire, it produces a magnetic field. If you are doing the activities in order, this should come as less-than-Earth-shattering news.

Motors spin because the magnetic field of the coil of wire is repelled by a second magnet that turns on and off. When it turns on and off, it attracts, then repels, then attracts, then repels the magnetic coil. This has the effect of pushing the coil around and around and around. When you switch wires, you are reversing the polarity (magnetism) of the coil—so what was attracted is now repelled and what was repelled is now attracted.

If this is simply all too confusing, may we politely but firmly recommend a copy of our magnet book, *Opposites Attract*. That, will hopefully clear a few things up.

Motor Effect

The Experiment

We are building up to more and more complex projects as we zip through the book. A motor is essentially a large coil of wire that spins in response to the presence of a magnetic field. This is what happens every time you jump in the car and turn over the ignition, start the blender, fire up the lawn mower, or use any other machine that incorporates a motor.

To better understand what causes the initial movement that gets these motors going, you are going to explore what is called the motor effect using a couple of magnets and a simple electrified wire.

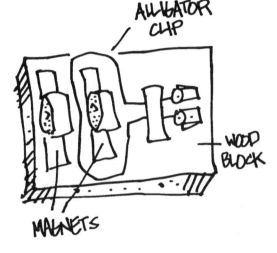

Materials

1   D Battery
1   D Battery clip
4   Donut magnets
1   Roll of masking tape
1   Wood board, 1 inch by 4 inches by 6 inches
3   Alligator clips
1   Compass

Procedure

1.   Insert the battery into the battery clip and attach one alligator lead to each terminal.

2.   Stack the four donut magnets into a pile and then divide them into two groups of two. Tape the magnets to the board, using the illustration to the upper right as a guide—north pole facing south pole leaving a gap of about one-half inch.

Motor Effect

3. Tape the third alligator clip in the position shown in the illustration below. The wire should be between the two sets of magnets. Tape the alligator clips to the board as shown.

4. Connect one of the leads from the battery to the alligator clip that is closest. Touch the other set of alligator clips to one another and observe what happens to the wire. Reverse the leads and see if the direction that the wire moves is different or remains the same.

How Come, Huh?

The four magnets produce a permanent magnetic field. When you connected the battery to the wire that was between the magnets, a second magnetic field surrounding the wire was produced. This magnetic field is going to be attracted to either the north pole (up) or the south pole (down) and will move in that direction. When you reverse the direction of the current in the wire, the magnetic field around the wire will also be reversed.

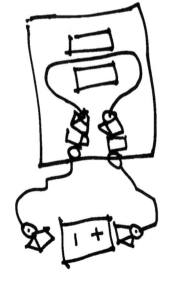

This all leads into one of the basic ideas of electricity that allows you to predict the direction of movement called the Right-Hand Rule.

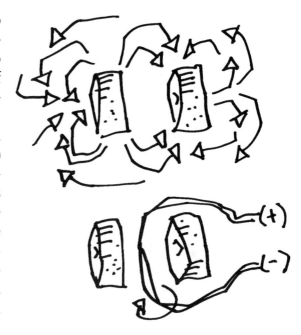

To use this rule you need to know two things, 1) the direction that the electricity is flowing (it leaves the battery at the negative terminal and re-enters the battery at the positive terminal), and 2) the orientation of the magnets (which can be determined using a compass).

Once you have determined the direction that the current is flowing, take your right hand and place it under the wire with your thumb sticking up. The wire should trace the position of your thumb like the drawing to the above right. Curl your fingers, forming a C. You can now predict the movement of the wire by looking at your fingers. The wire will either move up or down to mirror the direction that your fingers are pointing—another mystery of nature unraveled and revealed.

Science Fair Extensions

51. Change the diameter of the wire and see if the thickness has any effect on the rate or degree of movement.

52. Add or subtract magnets from the permanent magnetic field. Again, does the strength of the magnetic field affect, in any way, the rate or degree of movement in the wire.

53. Change the amount of electricity flowing through the wire by either adding batteries in series, increasing the voltage of the battery, or using a transformer with variable voltage.

A Simple Motor

The Experiment

We are definitely heading into the homestretch. Everything that you have been studying—all of the ideas that you have been poking and prodding—lead to your understanding of this one, big, idea: how a simple electric motor works.

We are going to use permanent magnets to create the magnetic field, a coil of wire to form the actual motor, and a felt marker will stand in for a decent commutator. Once you build this version, we will turn you loose on the model that I built over and over as a fifth grader in Mr. Goffard's class—but, first things, first.

Materials

1   D Battery holder
1   D Battery
4   Donut magnets
2   Alligator leads
1   12-oz. Plastic cup
1   Roll of masking tape
2   Paper clips, large
1   Pair of wire strippers
1   2-foot Length of bell wire
1   Black, felt marker, permanent

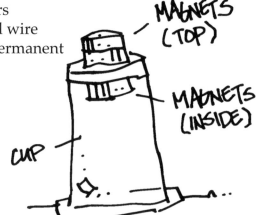

Procedure

1.   Insert the battery into the battery holder and attach one alligator lead to each terminal.

2. Make a stack out of the four donut magnets. Divide the stack in half and place two magnets inside the bottom of the cup and two additional magnets outside the bottom of the cup.

PAPER CLIP

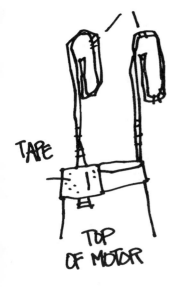

TAPE

TOP
OF MOTOR

3. Open the paper clips up and tape the extended portion of the paper clip to the bottom of the cup on either side of the magnets. Use the illustration to the left as a guide.

4. Leave a 2-inch tail and start to wrap a coil around two fingers on your hand. Wrap until all of the wire is used up, leaving another 2-inch tail.

Use the tails to make a single wrap around the coil and hold it in place. Adjust the tails so that they are sticking straight out. Strip two-thirds of the plastic off each tail.

5. Using a permanent marker, blacken the top half of each tail. Be sure to blacken the same side on both tails.

BLACKEN

COIL

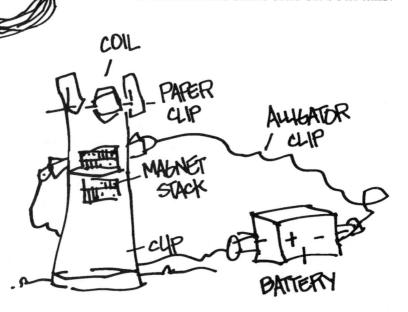

PAPER CLIP

ALLIGATOR CLIP

MAGNET STACK

CUP

BATTERY

A Simple Motor

6. Slide the coil of wire into the supports. Use the illustration on the bottom of the previous page to help you.

7. Clip the alligator leads from the battery to the base of each paper clip. When you connect the second lead, you will notice that the coil did one of three things, 1) started spinning, in which case you do nothing but holler, "Yahoo!" and shove your fist in the air, 2) wobbled, in which case give the coil a gentle twirl and it should start to spin, or 3) did nothing and you are going to have to troubleshoot (see below).

Troubleshooting

If you motor does not spin, try these things in this order.

1. Check your battery and make sure that it has juice.

2. Make sure that the alligator clips are connected to the metal tabs in the battery holder and not to a piece of the plastic.

3. Double-check the connection between the clip and the insulated wire; check for breaks in the alligator wire.

4. Straighten the tails coming from the coil. The coil should spin smoothly and without much wobbling when you spin it with your finger. If it does not, flatten the coil a bit and straighten the wire.

5. Check the position of the paper clips. They should be even. If one is higher than the other, the coil will be lopsided and have to work harder to spin. Make sure they are level.

6. Add a second battery for more juice or substitute either a lantern battery, 6-volt, or a variable, low-volt, power supply. By now your motor should be spinning. We have built thousands of these things and one of the reasons that we selected this design is because it is so kid- and adult-friendly—virtually any electrical nincompoop can build one.

After your motor spins for a while, you may notice that it starts working less efficiently. This is usually because the marker gets a little smeared. Clean the wire with a paper towel and re-mark the wire. Another common problem is that kids build these things, run them for hours, then they start to fritz. They readjust everything and still no spin. Get a new, fresh battery and this will solve your problem 99 percent of the time.

How Come, Huh?

The four magnets formed a permanent magnetic field. When you hooked the coil of wire to the battery, a magnetic field was produced around the coil. When these two magnetic fields come into close contact, there is movement. (If you just got here from the Motor Effect lab, this should sound very familiar.) The coil starts to move, but only temporarily.

It is temporary because you blackened the top half of each tail. This black ink served as an insulator so that when the coil makes half a spin, the electricity stops, the magnetic field disappears, but the coil continues to spin because it has momentum. When it completes its cycle, it comes in contact with the bare wire again, juice flows, magnetic fields are created, and the coil gets another push. Think of a bike that has been flipped upside down. You whack the wheel and it spins, you whack it again and it spins faster, you keep whacking the wheel each time it comes around and you have a lot of motion—same idea here.

Science Fair Extensions

54. Prove that the amount of electricity is directly proportional to the speed that the motor will spin.

55. Figure out other ways to insulate the top half of the wires that form the coil supports. Or, read up on commutators and figure out how to add one to your motor.

56. Experiment without insulating the coil supports.

Big Idea 10

Batteries are storage devices for electrons. They have a positive and a negative terminal and must be connected in an electrical loop for electrons to flow through a circuit.

Battery Lineup

– D

– C

– AA

– AAA

– WATCH

The Experiment

You have now been exposed to both kinds of electrical circuits— series and parallel. In both instances you added additional lightbulbs to the circuits and looked at the amount of light that the bulbs gave off.

In this lab you are going to double the juice, by adding a second battery, to both a series and parallel circuit and look at how that affects the brightness of the bulb.

Materials

1   Switch
6   Alligator clips
2   Batteries with battery clips
2   Bulbs with sockets

Procedure

1. Using the electrical diagram to the right, build the series circuit that is represented by the symbols. Close the switch and mentally record the brightness of the lightbulb.

2. Add a second battery and alligator clip to your series circuit using the bottom diagram as a guide. When they are in place, close the switch and mentally record the brightness of the lightbulb. Record your observations in the Data & Observations section.

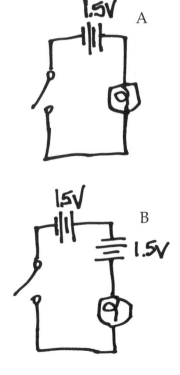

1.5V   A

1.5V   B

1.5V

Battery Lineup

3. Using the original electrical diagram on page 137, build the circuit that is represented by the symbols. Close the switch and mentally record the brightness of the lightbulb.

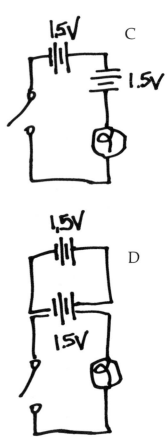

4. Add a second battery and alligator clips to your parallel circuit using the next diagram. When they are in place, close the switch and mentally record the brightness of the lightbulb. Record your observations in the Data & Observations section.

5. Finally, a combination of series and parallel. Use the diagram to the right and record your observations of the combination of the batteries and lightbulbs.

Data & Observations

Circle the answer that best describes your observations:

1. When the second battery was added to the series circuit, the amount of light given off in both bulbs (Diagram B) was _____ compared with just one bulb (Diagram A).

          A. dimmer   B. the same   C. brighter

2. When the second battery was added to the parallel circuit (Diagram C), the amount of light given off in both bulbs was _____ compared with just one bulb (Diagram A).

          A. dimmer   B. the same   C. brighter

3. When the second battery was added in parallel to the circuit (Diagram C), the amount of light given off by that bulb was _____ compared with amount of light in Diagram B.

      A. dimmer   B. the same   C. brighter

4. When the second battery was added in parallel to the series circuit (Diagram D), the amount of light given off in both bulbs was _____ compared with just one bulb (Diagram C).

      A. dimmer   B. the same   C. brighter

How Come, Huh?

In a parallel setup the electricity has choices as it flows through the circuit, but in a series circuit it does not. With respect to the different experiments you conducted, the answers that you should have come up with are:

1. In the first experiment, the bulbs got brighter. It is a series circuit; add another battery and you have more juice flowing through the wire—more juice, brighter bulb.

2. In the second experiment, everything is in parallel so adding a second battery should not have increased the brightness of the bulbs.

3. Batteries in parallel and bulbs in series: Bulbs get dimmer since they are sharing the electricity from the batteries.

4. Batteries in series and bulbs in parallel: Bulbs remain the same brightness.

Vinegar Voltage

The Experiment

Different metals have the ability to give or accept electrons. You are going to place two different metals, zinc and copper, in a dilute acid. The difference in these metals will allow electrons to flow from one to another through the weak acid.

Materials

3   Zinc strips, 0.25 inch by 4 inches
3   Copper strips, 0.25 inch by 4 inches
1   Lump of clay
3   250-ml Beakers
1   Bottle of distilled, white vinegar
4   Alligator clips
1   LED

Procedure

1. Insert a copper and zinc strip into one of the beakers. To keep the strips separated you can place a small lump of clay in the bottom of the beaker. Add 100 ml of vinegar to the cup and attach an alligator clip to each strip.

2. When you examine the LED (light emitting diode), you will notice that one wire is longer than the other. Connect the longer of the two wires to the alligator lead coming from the copper strip. Attach the shorter wire to the lead coming from the zinc strip. Look to see if any light is being produced—probably not.

Electron Herding 101 • B. K. Hixson

3. Insert zinc (abbreviated Zn) and copper (abbreviated Cu) strips in two other beakers. Using the diagram below, hook all three of the vinegar batteries in series. Be sure to always attach zinc to copper strips. When you get everything hooked up, see if the LED lights this time.

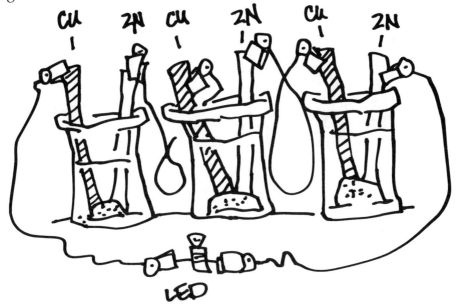

The zinc strip gives up electrons more easily than the copper strip, so when everything is hooked up in series, the electrons leave the zinc, zip through the vinegar—which gives the electrons a push and these electrons then enter the copper strip. The voltage, the number of electrons, is too low to light up the diode with just one beaker. When you hook three batteries in series—as you saw in the previous lab, the voltage increases and there is enough electricity to make the LED glow.

Science Fair Extensions

57. Repeat the experiment with other kinds of acids and see which one works the best as a battery. Be sure to get permission from your parents—some acids can be very dangerous.

Lemons in Series

The Experiment

Lemons make you pucker when you bite into them in part because of the citric acid that lemons, oranges, grapefruits, and kiwis all have.

If you think back to your liquid conductors lab, you will remember that water that had a little acid added to it was much more efficient at conducting electricity than just plain water. We are going to take it one more step with this lab. Not only are we going to conduct electricity but we are also going to make a battery that produces electricity using a lemon. What will those citrus growers think of next?

Materials

2   Lemons
1   Knife
4   Alligator clips
3   Zinc strips, 0.25 inch by 4 inches
3   Copper strips, 0.25 inch by 4 inches
1   LED
    Adult supervision/assistance

Procedure

1. You will want to get juicy, ripe lemons for this lab—the juicier the better. Take the two lemons and firmly roll them on the table to break up the tissue inside the lemon, but do not break the skin.

2. Have an adult help you cut both lemons in half. Set two lemon halves aside.

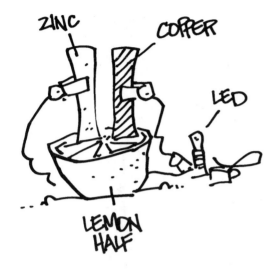

3. Insert a copper and zinc strip into the fleshy part of one half lemon. Avoid the white, pulpy rind. Attach an alligator clip to each strip.

4. When you look at the LED (light emitting diode), you will notice that one wire is longer than the other. Connect the longer of the two wires to the alligator lead coming from the copper strip. Attach the shorter wire to the lead coming from the zinc strip. Look to see if any light is being produced.

5. Insert zinc and copper strips in two of the other lemon halves. Using the diagram below hook two and then all three of the lemon halves in series. Be sure to always attach zinc to copper electrodes. When you get everything hooked up, see if the LED lights each time.

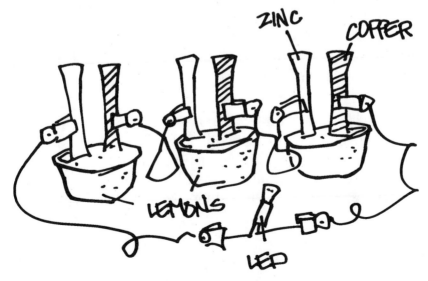

Lemons in Series

Data & Observations

Lemons	LED Y/N?
Single	
Double	
Triple	

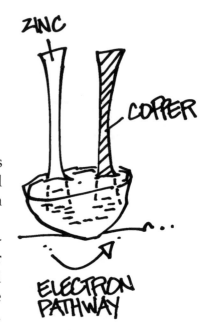

ZINC

COPPER

ELECTRON PATHWAY

How Come, Huh?

The juice in the lemon contains citric acid, which is a weak acid and also what is called an electrolyte—a solution that conducts electricity.

The zinc strip gives up electrons more easily than the copper strip, so when everything is hooked up in series the electrons leave the zinc, zip through the lemon juice, which gives it a push and enters the copper strip. The number of electrons—voltage—is too few to light up the diode. When you hook three batteries in series, as you saw in the previous lab, the voltage increases and with three lemon halves there is enough to make the LED glow.

Science Fair Extensions

58. Repeat the experiment with other kinds of citrus fruits and see which one works the best as a battery. You can buy voltmeters that are very sensitive and actually measure the voltage produced by lemons, oranges, grapefruits, and other citrus foods. Give that a try.

Hand Battery

The Experiment

You are going to generate an electric current that you are actually a component of by placing both hands on two dissimilar metals and letting the juice flow.

Materials

1 DC microammeter
  (100 microamps)
1 Aluminum plate, 6 inches by 6 inches
1 Copper plate, 6 inches by 6 inches
1 Lead plate, 6 inches by 6 inches
1 Zinc plate, 6 inches by 6 inches
2 Alligator clips
1 Towel

Procedure

1. Place both metal plates, copper and aluminum, on a nonmetallic surface.

2. Using the alligator clips, connect one plate to one of the meter's terminals and the other plate to the other terminal. Use the illustration to the right as a guide.

3. Place both hands on the plates and leave them there for 30 seconds. Check the reading on the microammeter over time.

If you are having difficulty getting a reading, rinse your hands under warm water, damp dry them on a towel, and place them on the plates again.

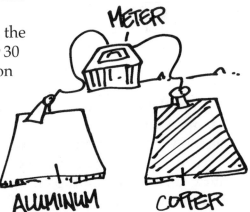

Hand Battery

4. Try the different combinations of metal plates suggested in the data table. Place your hands on the combinations of metals and wait about 30 seconds to take a reading.

Data & Observations

Plates	Ammeter Reading
Aluminum/Copper	
Aluminum/Lead	
Aluminum/Zinc	
Copper/Lead	
Copper/Zinc	
Lead/Zinc	

How Come, Huh?

When you touch the two plates, electrons flow from the copper plate into your hand. This leaves the copper plate with a positive charge. On the other plate is the opposite reaction. The aluminum grabs electrons from your hand and accumulates a negative charge. So, you have electrons flowing from the copper plate into one hand, through your body and out the other hand into the aluminum plate.

From the aluminum plate the excess electrons flow into the microammeter on their way to the copper plate. Depending on the combination of metals, you will get a variety of readings. The success you have using various metals will depend on a metal's electric potential, which is its ability to gain or lose electrons—the greater the difference between the two metals, the higher the voltage.

As an aside, the moisture produced by your hand acts like the battery acid in a battery. It lowers the resistance of movement to electrons on the surface of the skin. If the air is really dry, wetting your hands provides the necessary moisture.

Science Fair Extensions

59. Repeat the experiment with other kinds of metals and see which one works the best to produce the greatest potential.

ELECTRON FLOW

60. With the supervision of an adult, dissect a battery and examine the metals and electrolytes that are used.

Since most batteries are made using strong acids or alkaline substances you will want to wear goggles and gloves while you are examining these items.

Big Idea 11

Fuses protect circuits from over-heating and destroying valuable equipment.

Steel-Wool Fuses

The Experiment

This activity introduces the idea of a fuse. A fuse is an idea that was incorporated into electrical circuits to protect expensive appliances, stereos, lightbulbs, and most importantly, to prevent electrical fires.

Materials

1   Pad of steel wool, fine
4   Alligator clips
1   Batteries with battery clips
1   Bulb with socket (or buzzer)
1   Clock with second hand

Procedure

1. To make your fuses we are going to take fine steel wool and gently pull it apart to create individual strands of thin wire. These are your fuses.

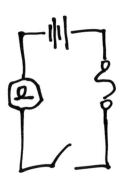

2. Using the illustration to the right, and the symbol for a fuse from above, construct the circuit pictured to the right. When everything else is ready, check the clock and connect your switch. Record the amount of time that it takes for the fuse to "blow." Repeat this a couple of times so that you can get an average.

3. Repeat this experiment using a double-stranded fuse, made by twisting two thin strands of steel wool together, and then use a triple-stranded fuse—yep, three strands twisted together. Test them and record the time that it takes for the fuse to blow.

Steel-Wool Fuses

Data & Observations

Fuse	Time to Blow
Single	
Double	
Triple	

How Come, Huh?

A fuse has one purpose and that is to protect the circuit. The way it does this is by disintegrating, or "blowing," when too much electricity is carried through the circuit. The fuse is designed to allow the proper amount of electricity to pass through the circuit. If there is a surge, or burst of energy, from the power plant, the same thing happens if the person who owns the house adds too many electrical things to the circuit and it is drawing too much electricity through the wires, the fuse gets hot and melts, again, the fuse saves the day. As soon as the fuse melts, the circuit is interrupted, the electricity stops flowing and appliances that were going to get too hot and burn up, or the wires that were going to catch on fire, have a chance to cool down.

Fuses are used in all kinds of things including stereos, cars, houses, large industrial machines, and anything else that needs to be protected from being overloaded.

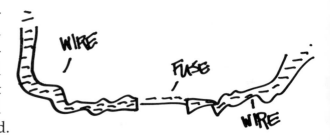

Science Fair Extensions

61.   Repeat the experiment but rather than changing the thickness of the fuse, change the strength of the battery.

Galvanometer

The Experiment

Hook a wire up to a battery full of electrons, and the electrons instantly zip into the wire. Not only do they deliver energy to the desired destination, but also a magnetic field is produced along the entire length of the wire. An electric current in a wire produces a magnetic field. You knew that. How about the reverse of that situation?

This lab will guide you in building an instrument called a galvanometer that will allow you to demonstrate that a magnetic field can also produce an electric current in a wire. The exact opposite of what you have been seeing? Why not.

Materials

1   Piece of cardboard, 3" by 4"
1   Roll of insulated bell wire
1   Pair of wire strippers
1   D Battery
1   D Battery clip
2   Alligator clips
1   Compass
1   Bar magnet

Procedure

1. Using the pattern on the next page, fold one-inch supports on each side of the cardboard soon to be a galvanometer.

2. Leave a 6-inch tail and wrap 30 wraps of wire around the middle of the galvanometer. Leave another 6-inch tail and cut the wire from the roll.

Galvanometer

3. Using the wire strippers, remove a half an inch of insulation off each tail, leaving nothing but bare wire exposed.

4. Snap the battery into the battery clip, and add an alligator clip to each side.

5. Slide the compass inside and under the wire that is around the piece of cardboard and see if the compass deflects (moves) at all. Rotate the compass and observe the position of the needle as you rotate it under the wire.

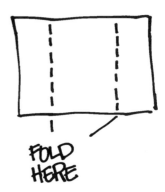

FOLD HERE

6. Rotate the piece of cardboard so that the compass needle is perpendicular to the loop of wire. See the illustration on the next page as a guide.

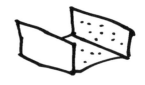

7. Hook one alligator clip up to each stripped end of the wire wrapped around the cardboard. Observe the position of the compass needle.

8. Congratulations! You have just built an instrument called a galvanometer. It is used by scientists to detect weak electric currents. But hang in there, you are only halfway through this lab. Now you are going to prove that a magnetic field can produce a weak electric current in a wire.

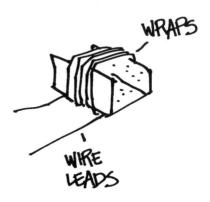

WRAPS

WIRE LEADS

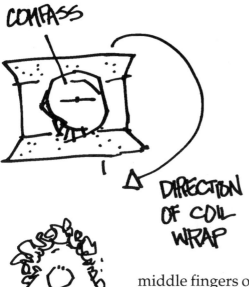

COMPASS

DIRECTION OF COIL WRAP

9. Disconnect the two alligator clips. The compass needle should have returned to its original position, which lets you know that a magnetic field is no longer present because an electric current is no longer flowing through the wire.

10. You are going to make another coil of wire. This time loosely wrap the wire around the three middle fingers of your left hand. Leave 6-inch tails at the beginning and end, just like you did with the galvanometer.

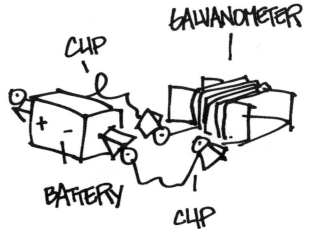

CLIP

GALVANOMETER

BATTERY

CLIP

Galvanometer

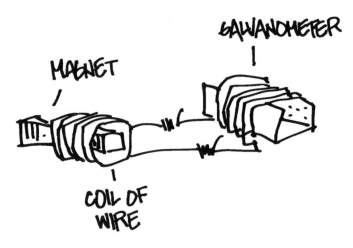

MAGNET

GALVANOMETER

COIL OF
WIRE

11. Strip the ends of the wire, then twist and connect the ends of the coil with the ends of the loop. This forms a giant circuit. Use the illustration at the top of the page as a guide.

12. Holding the loop in one hand and the magnet in the other, move the magnet in and out of the loop quickly. As you do, observe what happens to the compass needle.

How Come, Huh?

1. Following instruction 5, you placed the compass under the wire loop in the cardboard support. As you rotated the compass, the needle should have also rotated to always point to magnetic north. No current in the wire, no magnetic field, nothing to attract the needle in the compass. Three and out.

2. When you hooked the juice up to the wire loop, a magnetic field was produced. The way you know this is that the compass needle aligned with the wire loop. The only reason this would happen is if a magnetic field suddenly appeared to wrangle the needle into position.

In summary, you hook the wire loop to a battery, the electricity starts to flow, the electrons flowing through the wire produce a weak magnetic field that attracts the compass needle. Unhook the battery and the compass needle rotates back to its original position.

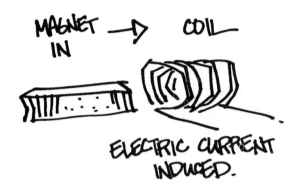

MAGNET IN → COIL

ELECTRIC CURRENT INDUCED.

3. When you inserted the magnet in the coil loop, the magnetic field of that magnet induced, or started, the electrons in the wire flowing. The moving electrons are defined as a current, and an electric current flowing through a wire produces a magnetic field. The magnetic field in the wire attracted the compass needle, and you now have proof positive that a magnet can create a weak electric current. You are also a blossoming expert on galvanometer construction.

Science Fair Extensions

62. Design an experiment that allows you to test the effect of increasing or decreasing the number of loops in the coil on the strength of the magnetic field.

63. Double up the bar magnet you use or substitute a stronger magnet, like a cow magnet, and determine how that affects the outcome of your experiment.

64. Use your galvanometer to demonstrate that the magnetic field in a coil of wire is reversed when the leads connecting the coil to the battery are reversed.

Electric Potato Pie

The Experiment

We have electrocuted lemons and chased electrons through vinegar. We have produced sound, light, heat, magnetic effects, and even coated metals with metals using electricity. Why not run some juice through a potato to see what happens?

Materials

1   Potato, large, fresh
1   Knife
2   Copper wires, insulated, 4 inches each
1   Pair of wire strippers
2   Alligator clips
1   Transistor battery, 9 volt
    Adult supervision

Procedure

1. Slice the potato in half with the knife.

2. Using the wire strippers remove the plastic coating from both ends of each piece of the bell wire.

3. Insert the stripped ends of the wires into the potato, spacing them about 1.5 inches apart. Use the illustration to the right as a guide.

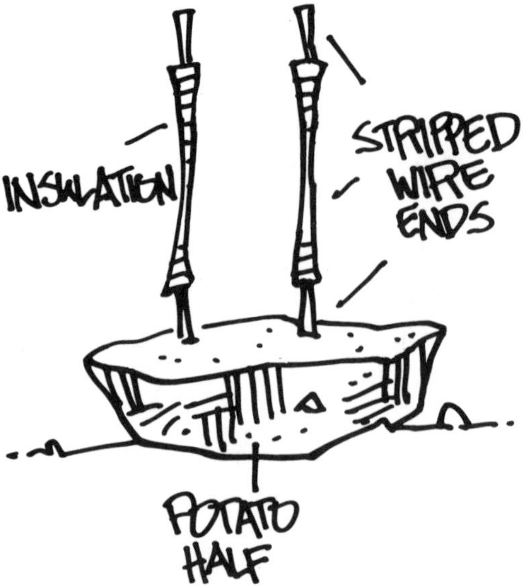

INSULATION

STRIPPED WIRE ENDS

POTATO HALF

4. Attach one alligator clip to the other end of each wire that is sticking out of the potato. Connect the other ends of the alligator clips to the 9-volt battery.

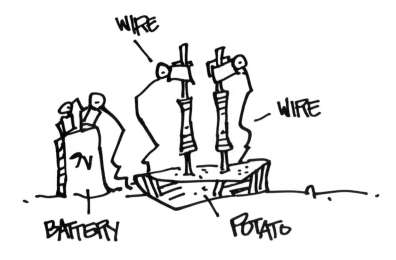

WIRE

WIRE

7V

BATTERY

POTATO

Data & Observations

You will notice that there is a reaction at the end of each terminal. Take a peek and record what you see.

Terminal	Reaction
Positive	
Negative	

How Come, Huh?

The electrons flowing through the potato caused a reaction at both terminals. The negative terminal produced hydrogen bubbles where the electricity caused the water in the potato to separate into hydrogen and oxygen. The positive terminal turned green where copper ions from the wire migrated out into the potato tissue.

Science Fair Extensions

65. Try other roots. Yams, sweet potatoes, turnips, radishes, carrots, onions, among others, are all likely candidates for extended experiments.

Dimmer Primer

The Experiment

We haven't really gotten into the idea of resistance too much in this book, we were planning on saving it more for the electronics book in the series, *The Adventures of Diode Dude*. However, in the spirit of Hollywood, here is a sneak peek at an idea that is very important when it comes to both electricity and electronics—resistance.

Materials

1   Pencil
1   Kitchen knife, sharp
3   Alligator clips
1   Bulb and socket
1   D Battery and clip
     Adult assistance

Procedure

1. Ask an adult to score and remove the top half of a sharpened pencil with the kitchen knife. This will expose the graphite shaft (pencil lead) that you write with.

2. Connect the alligator clips to the battery terminals on the holder and insert the lightbulb where it is shown in the series to the right.

3. Finally clip one of the alligator clips to the sharpened pencil point and lay the other alligator clip on the graphite shaft of the pencil.

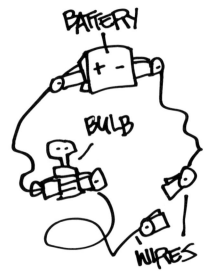

4. Slowly move the second alligator clip along the graphite shaft of the pencil toward the alligator clip attached to the point of the pencil. Observe the brightness of the lightbulb as you move the alligator clip.

Data & Observations

Distance between Clip	Brightness of Bulb
8 inches	
6 inches	
4 inches	
2 inches	
0.01 inches	

How Come, Huh?

The graphite resists, or impedes, the flow of electrons. The farther apart the clips are, the dimmer the bulb appears because fewer electrons make it all the way through the pencil shaft.

Science Fair Extensions

66. Replace the pencil with real resistors.

Quiz Board

The Experiment

You can make a fun electrical game, called a quiz board, using a simple circuit and a bunch of conductors. It consists of two columns on the front of the board matched by wires on the back. You match the question with the answer by touching them both at the same time with a pair of alligator clips. If you are correct, a light or buzzer sounds; if you are incorrect, there is no sound.

Materials

1   Sheet of cardboard, 8 inches by 11 inches
1   Pen or pencil
1   Ruler
1   Box of brads
1   Roll of bell wire
1   Pair of wire strippers
3   Alligator clips
1   D Battery with clip
1   Buzzer or lamp with socket

Procedure

1. Place the ruler along the edge of the cardboard and make a line one inch from the edge. Do the same thing for the other side.

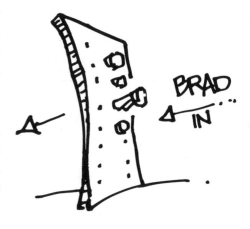

2. Starting one inch from the top of the board mark a dot every two inches along both lines. Use the illustration to the right as a guide.

3. Shove a brad into the board on top of every dot and fold the back side of the brad on to the back.

4. Cut 5 pieces of bell wire. Each piece should be at least 12 inches long. Strip both ends of all wires.

5. Write the names of five states on the front of the board. Write the names of the state capitals on the other side of the board—but place them out of order. If you use pieces of paper instead of writing directly on the board, you can use the board over and over.

6. On the back of the quiz board connect the state to the correct state capital by running a stripped wire between the two brads and wiring them in. See the illustration above. Match each state with its correct state capital and turn the board over.

7. Connect the alligator clips to both halves of the battery clip and then connect the buzzer to one of the alligator clips and a second alligator clip to the other side of the buzzer.

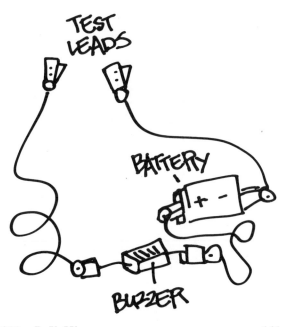

Quiz Board

8. Some buzzers are fickle so complete the circuit to make sure that the buzzer is going to make a sound. If is does not, reverse the connections.

9. When you are ready to go, touch one alligator clip to a state and touch the second alligator clip to the city that you think is the capital of that state. If you are correct, the buzzer will sound; if you are off the mark, there will be a telling silence.

Science Fair Extensions

67. There was a very popular quiz game called *College Bowl* in the 1970s, and it has been resurrected every couple of years. It consists of two teams of four players. Each player has a button they can hit. The host asks questions, and the first person to hit the buzzer lights up a light on their side of the table. Construct the electric circuit that would allow you to play this game with your friends.

Exploding Water

The Experiment

Electricity can also be used to manufacture products. In this case we are going to introduce you a process called hydrolysis that uses electricity to split water molecules into hydrogen and oxygen gases. Both of these gases are sold commercially and have different characteristics that you will explore in this lab.

Materials

1   D battery holder
1   D battery
1   1 oz. bottle of phenolphthalein
     in ethyl alcohol
1   9 oz. clear, plastic, tumbler
1   1 oz. bottle of sodium sulfate
1   Craft stick
1   Plastic test tube
1   Cup
1   Insulated copper wire
1   Bare copper wire
2   Alligator clips
1   Book of matches
    Water
    Adult Supervision

Procedure

1. Insert the D battery in the battery holder. Connect one alligator clip to each battery terminal and set this apparatus aside for a moment.

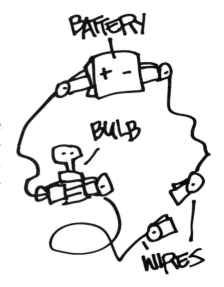

Exploding Water

2. Fill the 9 oz. plastic cup with water and add one capful of the phenolphthalein. Now add one capful of sodium sulfate powder to the mixture and stir the contents of the cup with the craft stick until all of the powder has dissolved.

3. Fill the plastic test tube with the solution you've just made. Place your thumb over the top of the test tube, turn it upside down, and place it in the cup. Lean the test tube up against the wall of the cup and locate the insulated wire. It should have two bare, exposed ends.

4. Slide the insulated wire underneath the mouth of the test tube, so that one end of the wire is at least halfway into the tube, but still submerged. Drape the remaining wire over the top of the cup, so it can be connected to the battery later.

5. Place one end of the bare copper wire in the solution. Drape the remainder of the wire over the side so it can also be connected to the battery.

6. Using the alligator clips, connect the insulated wire to the negative side of the battery, and the bare copper wire to the positive terminal. Observe the color changes that take place inside the test tube as the electricity starts to flow.

7. Allow the gas to collect inside the tube for a few minutes, then carefully lift the tube up, keeping it upside down. When an adult comes to your lab area, place your thumb over the end of the tube and flip the tube right side up. An adult will light a match and bring it near the edge of the test tube. When the match gets close, quickly remove your thumb and watch the reaction.

How Come, Huh?

Water is composed of 2 parts hydrogen hooked to one part oxygen. As a general rule the atoms stay hooked together until something produces enough energy to split them apart. In this case the electricity is the energy source splitting the atoms.

Hydrogen is flammable. It also has a positive charge. That means that if hydrogen gas is produced when an electric current travels through water the hydrogen will collect at the negative terminal of the current. When the match was brought near the hydrogen gas, it ignited and burned rapidly, expanding and causing a whoop when the air rushed out of the tube.

The pink color is produced by phenolpthalein, an acid/base indicator. As the reaction proceeded the pH, the number of free ions in the solution changed, which was reflected by the color change.

Homemade Flashlight

The Experiment

This is the culminating activity for the entire book. We are going to take stored electrical energy—a battery—and construct a device to convert that electrical energy to light energy—a flashlight. Along the way you will assemble another homemade switch to control the flow of electrons and a complete circuit to carry the electrons.

Materials

2   35 mm film canisters, empty
2   Thumbtacks, metal
4   Brads, brass
2   C Batteries
1   Christmas tree bulb
4   Pieces of bell wire, 3 inch (1), 5 inch (1), 7 inch (2)
1   Pair of wire strippers
1   Piece of wood, 2 inches by 0.25 inch by 0.25 inch
1   Paper clip, small
1   Piece of thin tagboard, 4 inches by 4inches
1   Pencil
1   Pair of scissors
1   Paper-hole punch
1   Roll of transparent tape
1   Piece of aluminum foil, 4 inches by 4inches
1   6-inch length of 1.5inch PVC pipe, with two holes
1   1.5 inch PVC cap
1   Roll of electrical tape
1   Piece of plastic wrap or cellophane, 4 inches by 4 inches
1   PVC reducer, 2 inches to 1.5 inch
1   Permanent marker

Procedure

Making the battery cases . . .

1. Punch a brass brad into the center of each end of the plastic camera film canisters—bottom and lid. Open the brads after they are inserted. These are your battery contacts.

2. Once the brads are in place, insert the batteries inside the camera film canisters, postive end up (the end with the little bump). Snap on the lids. Touch the leads from the Christmas bulb to each brad on the end of the film container to make sure there is contact being made inside the can.

3. Wrap one end of the 3-inch bell wire to the brad at the bottom of one battery case and then connect the two battery holders by wrapping the other end of the wire to the brad at the top of the second battery case. (Wrapping the wire around the brads may be easier if the batteries are taken out first.)

Making the switch . . .

1. Using the wire strippers, remove an inch of the plastic insulation from both ends of both 7-inch wires. If you are not sure how to use the wire strippers, please ask an adult for assistance.

2. Once the wires are stripped, wrap one end of the first 7-inch wire around one thumbtack and press it into the wood block. Use the illustration on page 168 as a guide.

Homemade Flashlight

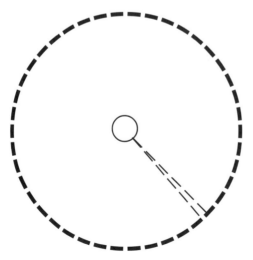

3. Using the second 7-inch wire, wrap one stripped end to the second thumbtack. Slide one end of the metal paper clip over the point of the thumbtack and then press it into the wooden rectangle about an inch from the first thumbtack. Make sure that the paper clip can rotate back and forth, touching the other thumbtack.

Making the flashlight bulb reflector shield . . .

1. Copy the dashed circle pattern pictured below on to the 4-inch by 4-inch square of tagboard and cut it out using the scissors. Draw a line directly across to the middle of the circle, similar to the picture below. This will be called your radial line.

2. Start at the edge of the circle and cut through the tagboard on the radial line to the center.

3. Using a handheld hole punch, punch a hole in the center of the tagboard. The tagboard should now look like the disc below.

4. Wrap the circle with aluminum foil, making sure the shiny side is facing outward. Fold all excess aluminum foil over and onto the back of the reflector, and tape it in place with the transparent tape. Smooth the aluminum foil on the inside of the reflector so that it has as few wrinkles as possible.

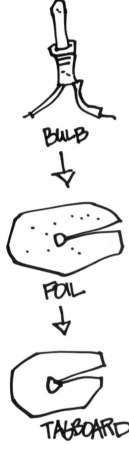

BULB

↓

FOIL

↓

TAGBOARD

5. Using the tip of the pencil, push through the aluminum foil. Shape the reflector so that it has a concave shape (it bends in a bit). If the ends of the Christmas bulb holder have not been stripped, now would be a good time.

6. Insert the ends of the wires through the holes from the shiny, front side of the reflector. Gently push the lightbulb into the center hole of the reflector shield so that it is securely held in place.

Assembling the flashlight . . .

1. Locate the large PVC cylinder. You will notice that there are two small holes in the side of the cylinder. Place the switch between the holes and insert each of the bell wires that are attached to the switch through the drilled holes in the PVC tube. String them out of each end of the PVC tube. Use the illustration on page 170 as a guide.

2. Wrap one of the switch wires to one of the wires on the lightbulb and then cover this connection with a small piece of electrical tape. This is the top end of your flashlight.

3. Wrap the other switch wire to the brad at the bottom of the battery case pair. This is the bottom end of your flashlight.

4. Wrap one end of the 5-inch wire to the brad at the top of the remaining battery case.

Homemade Flashlight

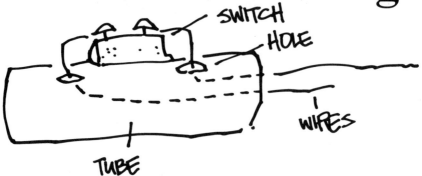

5. Slide the battery cases inside the PVC tube. Place the cap on the bottom of the tube. Use the illustration on page 171 as a guide.

6. Wrap the other end of the 5-inch wire to the other flashlight bulb wire and cover this connection with electrical tape.

7. Put the reflector shield at the top of the flashlight tube, pressing it slightly concave so that it is almost secure. Cover with the 4-inch by 4-inch piece of plastic wrap. Push the PVC reducer onto it to hold it all firmly in place. Trim plastic wrap with scissors.

8. Electrical tape the on/off switch in place. To personalize your flashlight you may decorate your construction with markers and or stickers. Be sure to write your name or initials on the bottom of the flashlight.who

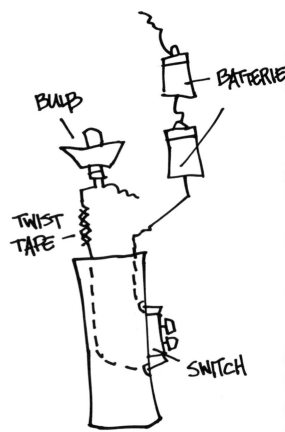

9. To test your flashlight rotate the paper clip, connecting both thumbtacks. If all of the connections have been made properly the bulb will light and you may now holler, "Yahoo!"

Data & Observations

In the space provided below draw a picture of the electrical circuit that you just built. Be sure to use the correct symbols and connect everything together in a complete circuit.

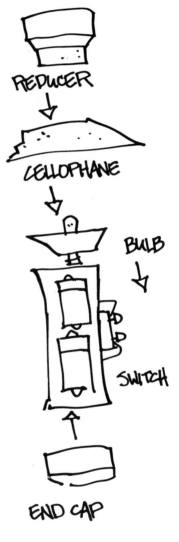

REDUCER

CELLOPHANE

BULB

SWITCH

END CAP

How Come, Huh?

When you placed the metal brads in the ends of the film containers, you created a pathway, through the metal conductors, for the electricity to flow. By connecting the two batteries in series, using the wire from the top of one container and connecting it to the brad at the bottom of the other container you "doubled the juice."

The paper clip was also a conductor and doubled as a switch— one wire from the switch was connected to the bottom of the battery stack. The two batteries are in series. The top of the battery stack was connected to one of the wires in the bulb. The other wire in the bulb was connected to the other end of the switch. So, when you connected the switch and completed the circuit, the electrons in the batteries zipped up the wire into the bulb, through the bulb, out the bulb, into the switch, through the switch, out the switch and back into the battery. Science is so cool.

That's it, you've made it to the end of the book, and if you explored along the way, your universe is richer and more diverse than when you started. Join us for another book when you get the chance.

Science Fair Projects
•
A Step-by-Step Guide: From Idea to Presentation

Electron Herding 101 • © 2002 • B. K Hixson

Science Fair Projects

Ah, the impending science fair project. A good science fair project has the following five characteristics:

1. The student must come up with an *original* question.

2. That *original* question must be suited to an experiment in order to provide an answer.

3. The *original* idea is outlined with just one variable isolated.

4. The *original* experiment is performed and documented using the scientific method.

5. A presentation of the *original* idea in the form of a lab write-up and display board is completed.

Science Fair Projects

As simple as science fair versus science project sounds, it gets screwed up millions of times a year by sweet, unsuspecting students who are counseled by sweet, unknowing, and probably just as confused parents.

To give you a sense of contrast we have provided a list of legitimate science fair projects and then reports that do not qualify. We will also add some comments in italics that should help clarify why they do or do not qualify in the science fair project department.

Science Fair Projects

1. Temperature and the amount of time it takes mealworms to change to beetles.

Great start. We have chosen a single variable that is easy to measure: temperature. From this point forward the student can read, explore, and formulate an original question that is the foundation for the project.

A colleague of mine actually did a similar type of experiment for his master's degree. His topic: The rate of development of fly larva in cow poop as a function of temperature. No kidding. He found out that the warmer the temperature of the poop the faster the larva developed into flies.

2. The effect of different concentrations of soapy water on seed germination.

Again, wonderful. Measuring the concentration of soapy water. This leads naturally into original questions and a good project.

3. Crystal size and the amount of sugar in the solution.

This could lead into other factors such as exploring the temperature of the solution, the size of the solution container, and other variables that may affect crystal growth. Opens a lot of doors.

vs. Science Reports

4. Helicopter rotor size and the speed at which it falls.

Size also means surface area, which is very easy to measure. The student who did this not only found the mathematical threshold with relationship to air friction, but she had a ton of fun.

5. The ideal ratio of baking soda to vinegar to make a fire extinguisher.

Another great start. Easy to measure and track, leads to a logical question that can either be supported or refuted with the data.

Each of those topics *measures* one thing such as the amount of sugar, the concentration of soapy water, or the ideal size. If you start with an idea that allows

you to measure something, then you can change it, ask questions, explore, and ultimately make a *prediction*, also called a *hypothesis*, and experiment to find out if you are correct. Here are some well-meaning but misguided entries:

Science Reports, <u>not Projects</u>
1. Dinosaurs!

OK, great. Everyone loves dinosaurs but where is the experiment? Did you find a new dinosaur? Is Jurassic Park alive and well, and we are headed there to breed, drug, or in some way test them? Probably not. This was a report on T. rex. Cool, but not a science fair project. And judging by the protest that this kid's mom put up when the kid didn't get his usual "A", it is a safe bet that she put a lot of time in and shared in the disappointment.

More Reports &

2. Our Friend the Sun

Another very large topic, no pun intended. This could be a great topic. Sunlight is fascinating. It can be split, polarized, reflected, refracted, measured, collected, converted. However, this poor kid simply chose to write about the size of the sun, regurgitate facts about its features, cycles, and other astrofacts while simultaneously offending the American Melanoma Survivors Society. Just kidding about that last part.

3. Smokers' Poll

A lot of folks think that they are headed in the right direction here. Again, it depends on how the kid attacks the idea. Are they going to single out race? Heredity? Shoe size? What exactly are they after here? The young lady who did this report chose to make it more of a psychology-studies effort than a scientific report. She wanted to know family income, if they fought with their parents, how much stress was on the job, and so on. All legitimate concerns but not placed in the right slot.

4. The Majestic Moose

If you went out and caught the moose, drugged it to see the side effects for disease control, or even mated it with an elk to determine if you could create an animal that would become the spokesanimal for the Alabama Dairy Farmers' Got Melk? promotion, that would be fine. But, another fact-filled report should be filed with the English teacher.

5. How Tadpoles Change into Frogs

Great start, but they forgot to finish the statement. We know how tadpoles change into frogs. What we don't know is how tadpoles change into frogs if they are in an altered environment, if they are hatched out of cycle, if they are stuck under the tire of an off-road vehicle blatantly driving through a protected wetland area. That's what we want to know. How tadpoles change into frogs, if, when, or under what measurable circumstances.

Now that we have beat the chicken squat out of this introduction, we are going to show you how to pick a topic that can be adapted to become a successful science fair project after one more thought.

One Final Comment

A Gentle Reminder

Quite often I discuss the scientific method with moms and dads, teachers and kids, and get the impression that, according to their understanding, there is one, and only one, scientific method. This is not necessarily true. There are lots of ways to investigate the world we live in and on.

Paleontologists dig up dead animals and plants but have no way to conduct experiments on them. They're dead. Albert Einstein, the most famous scientist of the last century and probably on everybody's starting five of all time, never did experiments. He was a theoretical physicist, which means that he came up with a hypothesis, skipped over collecting materials for things like black holes and space-time continuums, didn't experiment on anything or even collect data. He just went straight from hypothesis to conclusion, and he's still considered part of the scientific community. You'll probably follow the six steps we outline but keep an open mind.

Project Planner

This outline is designed to give you a specific set of time lines to follow as you develop your science fair project. Most teachers will give you 8 to 11 weeks notice for this kind of assignment. We are going to operate from the shorter time line with our suggested schedule, which means that the first thing you need to do is get a calendar.

A. The suggested time to be devoted to each item is listed in parentheses next to that item. Enter the date of the Science Fair and then, using the calendar, work backward entering dates.

B. As you complete each item, enter the date that you completed it in the column between the goal (due date) and project item.

Goal   *Completed*   *Project Item*

1. Generate a Hypothesis (2 weeks)

Goal	Completed	Project Item
_____	_____	Review Idea Section, pp. 182–190
_____	_____	Try Several Experiments
_____	_____	Hypothesis Generated
_____	_____	Finished Hypothesis Submitted
_____	_____	Hypothesis Approved

2. Gather Background Information (1 week)

Goal	Completed	Project Item
_____	_____	Concepts/Discoveries Written Up
_____	_____	Vocabulary/Glossary Completed
_____	_____	Famous Scientists in Field

& Timeline

Goal   Completed   Project Item

3. Design an Experiment (1 week)

Goal	Completed	Project Item
_____	_____	Procedure Written
_____	_____	Lab Safety Review Completed
_____	_____	Procedure Approved
_____	_____	Data Tables Prepared
_____	_____	Materials List Completed
_____	_____	Materials Acquired

4. Perform the Experiment (2 weeks)

_____	_____	Scheduled Lab Time

5. Collect and Record Experimental Data (part of 4)

_____	_____	Data Tables Completed
_____	_____	Graphs Completed
_____	_____	Other Data Collected and Prepared

6. Present Your Findings (2 weeks)

_____	_____	Rough Draft of Paper Completed
_____	_____	Proofreading Completed
_____	_____	Final Report Completed
_____	_____	Display Completed
_____	_____	Oral Report Outlined on Index Cards
_____	_____	Practice Presentation of Oral Report
_____	_____	Oral Report Presentation
_____	_____	Science Fair Setup
_____	_____	Show Time!

Scientific Method
• Step 1 •
The Hypothesis

Electron Herding 101 • © 2002 • B. K Hixson

The Hypothesis

A hypothesis is an educated guess. It is a statement of what you think will probably happen. It is also the most important part of your science fair project because it directs the entire process. It determines what you study, the materials you will need, and how the experiment will be designed, carried out, and evaluated. Needless to say, you need to put some thought into this part.

There are four steps to generating a hypothesis:

Step One • Pick a Topic
Preferably something that you are interested in studying. We would like to politely recommend that you take a peek at physical science ideas (physics and chemistry) if you are a rookie and this is one of your first shots at a science fair project. These kinds of lab ideas allow you to repeat experiments quickly. There is a lot of data that can be collected, and there is a huge variety to choose from.

If you are having trouble finding an idea, all you have to do is pick up a compilation of science activities (like this one) and start thumbing through it. Go to the local library or head to a bookstore and you will find a wide and ever-changing selection to choose from. Find a topic that interests you and start reading. At some point an idea will catch your eye, and you will be off to the races.

Pick An Idea You Like

We hope you find an idea you like between the covers of this book. But we also realize that 1) there are more ideas about electricity than we have included in this book, and 2) other kinds of presentations, or methods of writing labs, may be just what you need to trigger a new idea or put a different spin on things. So, without further adieu, we introduce you to several additional titles that may be of help to you in developing a science fair project.

For Kids of All Ages . . .

1. The Cool Hot Rod and Other Electrifying Experiments on Energy & Mattter. Written by Paul Doherty, Don Rathjen, and the Exploratorium Teacher Institute. ISBN 0-471-11518-5 Published by John Wiley & Sons, Inc. 100 pages.

This is a another wonderful book from our favorite hands-on science center, the Exploratorium in San Francisco, California. It contains twenty-two science "snacks." These are mini-versions of the larger exhibits that you can find at the Exploratorium. Each snack has been designed by classroom teachers and is kid-tested. In addition to sections giving an overview, a list of materials, assembly instructions (with estimated times), things to do and notice, and questions about what is going on, there is also an etc. section that will give you ideas for developing the lab further. The book also comes with excellent illustrations and photographs to guide you in the construction of the lab.

2. Janice Van Cleave's Electricity. Written by Janice Van Cleave. ISBN 0-471-31010-7 Published by John Wiley & Sons. 90 pages.

There are 20 very well thought-out activities in this book. In addition to laying out the lab in a clear and easy-to-follow method, she has also devoted much more time to expanding on the idea and following up with extensions—in particular, science fair ideas. Laced with historical references and good illustrations.

3. *Electricity & Magnetism FUNdamentals. FUN tastic Science Activities for Kids. Written by Robert W. Wood. ISBN 0-7910-4841-1 Published by Chelsea House. 132 pages.*

As the title implies, this book is split half and half between magnets and electrical circuits, and it does a very good job of presenting important ideas for both topics. In the electricity department there are fifteen great labs. Each lab comes with clear instructions, good illustrations, and most of the labs are tagged with a couple fun trivia ideas.

4. *Science Experiments with Electricity. Written by Sally Nankevellston and Dorothy Jackson ISBN 0-531-15442-2 Published by Franklin Watts. 32 pages.*

This is the most kid-friendly book of the six that we are reviewing. The text is written as if the author were there talking with you. It has cute titles, well-done illustrations, concise directions, and easy-to-understand explanations. The book also covers a wide range of basic ideas on electricity.

5. *The Science Book of Electricity. Written by Neil Ardley ISBN 0-15-200583-8 Published by Harcourt, Brace, Jovanovich 32 pages.*

Also a very kid-friendly book. Photographs of each step of the way guide you through thirteen lab activities that will give you a good overview of the principles of electricity.

6. *Electricity. Written by Steve Parker. ISBN 0-7894-6169-2 Published by Dorling Kindersley. 64 pages.*

We included this book for a historical perspective of the development of the science more than for the experiments, which are described but not put in a replicable form for you. Excellent background on the scientists and the ideas from the earliest discoveries right on up through today.

Develop an Original Idea

Step Two • Do the Lab

Choose a lab activity that looks interesting and try the experiment. Some kids make the mistake of thinking that all you have to do is find a lab in a book, repeat the lab, and you are on the gravy train with biscuit wheels. Your goal is to ask an ORIGINAL question, not repeat an experiment that has been done a bazillion times before.

As you do the lab, be thinking not only about the data you are collecting, but of ways you could adapt or change the experiment to find out new information. The point of the science fair project is to have you become an actual scientist and contribute a little bit of new knowledge to the world.

You know that they don't pay all of those engineers good money to sit around and repeat other people's lab work. The company wants new ideas so if you are able to generate and explore new ideas you become very valuable, not only to that company but to society. It is the question-askers that find cures for diseases, create new materials, figure out ways to make existing machines energy efficient, and change the way that we live. For the purpose of illustration, we are going to take a lab from this book and run it through the rest of the process. Turn to page 52 and the lab titled, " Prisms, Water Prisms." The lab uses a tub of water, an ordinary mirror, and light to create a prism that splits the light into the spectrum of a rainbow. Cool. Easy to do. Not expensive and open to all kinds of adaptations, including the four that we provide in the Science Fair Extension section on page 55.

Step Three • *Bend, Fold, Spindle, & Mutilate Your Lab*

Once you have picked out an experiment, ask if it is possible to do any of the following things to modify it into an original experiment. You want to try and change the experiment to make it more interesting and find out one new, small piece of information.

Heat it	Freeze it	Reverse it	Double it
Bend it	Invert it	Poison it	Dehydrate it
Drown it	Stretch it	Fold it	Ignite it
Split it	Irradiate it	Oxidize it	Reduce it
Chill it	Speed it up	Color it	Grease it
Expand it	Substitute it	Remove it	Slow it down

If you take a look at page 55, that's exactly what we did to the main idea. We took the list of 24 different things that you could do to an experiment—not nearly all of them by the way—and tried a couple of them out on the prism setup.

Double it: 20. Get a second prism and see if you can continue to separate the colors farther by lining up a second prism in the rainbow of the first.

Reduce it: 21. Figure out a way to gather up the colors that have been produced and mix them back together to produce white light again.

Reverse it: 22. Experiment with moving the flashlight and paper closer to the mirror and farther away. Draw a picture and be able to predict what happens to the size and clarity of the rainbow image.

Substitute it: 23. You can also create a rainbow on a sunny day using a garden hose with a fine-spray nozzle attached. Set the nozzle adjustment so that a fine mist is produced and move the mist around in the sunshine until you see the rainbow. This works better if the sun is lower in the sky; late afternoon is best.

Hypothesis Work Sheet

Step Three (Expanded) • *Bend, Fold, Spindle Work Sheet*
This work sheet will give you an opportunity to work through the process of creating an original idea.

A. Write down the lab idea that you want to mangle.

_____

_____

_____

_____

B. List the possible variables you could change in the lab.

    i. _____

    ii. _____

    iii. _____

    iv. _____

    v. _____

C. Take one variable listed in section B and apply one of the 24 changes listed below to it. Write that change down and state your new lab idea in the space below. Do that with three more changes.

Heat it	Freeze it	Reverse it	Double it
Bend it	Invert it	Poison it	Dehydrate it
Drown it	Stretch it	Fold it	Ignite it
Split it	Irradiate it	Oxidize it	Reduce it
Chill it	Speed it up	Color it	Grease it
Expand it	Substitute it	Remove it	Slow it down

    i. _____

_____

_____

_____

ii. _____

_____

_____

_____

iii. _____

_____

_____

_____

iv. _____

STRETCHING!

_____

_____

Step Four • Create an Original Idea— Your Hypothesis

Your hypothesis should be stated as an opinion. You've done the basic experiment, you've made observations, you're not stupid. Put two and two together and make a PREDICTION. Be sure that you are experimenting with just a single variable.

D. State your hypothesis in the space below. List the variable.

i. _____

_____

_____

_____

ii. Variable tested: _____

Sample Hypothesis Work Sheet

On the previous two pages is a work sheet that will help you develop your thoughts and a hypothesis. Here is sample of the finished product to help you understand how to use it.

A. Write down the lab idea that you want to mutilate.
A mirror is placed in a tub of water. A beam of light is focused through the water onto the mirror, producing a rainbow on the wall.

B. List the possible variables you could change in the lab.
    i. **Source of light**
    ii. **The liquid in the tub**
    iii. **The distance from flashlight to mirror**

C. Take one variable listed in section B and apply one of the 24 changes to it. Write that change down and state your new lab idea in the space below.

The shape of the beam of light can be controlled by making and placing cardboard filters over the end of the flashlight. Various shapes such as circles, squares, and slits will produce different quality rainbows.

D. State your hypothesis in the space below. List the variable. Be sure that when you write the hypothesis you are stating an idea and not asking a question.

Hypothesis: The narrower the beam of light the tighter, brighter, and more focused the reflected rainbow will appear.

Variable tested: **The opening on the filter**

Scientific Method
• Step 2 •
Gather Information

Gather Information

Read about your topic and find out what we already know. Check books, videos, the Internet, and movies, talk with experts in the field, and molest an encyclopedia or two. Gather as much information as you can before you begin planning your experiment.

In particular, there are several things that you will want to pay special attention to and that should accompany any good science fair project.

A. Major Scientific Concepts
Be sure that you research and explain the main idea(s) that is / are driving your experiment. It may be a law of physics or chemical rule or an explanation of an aspect of plant physiology.

B. Scientific Words
As you use scientific terms in your paper, you should also define them in the margins of the paper or in a glossary at the end of the report. You cannot assume that everyone knows about geothermal energy transmutation in sulfur-loving bacterium. Be prepared to define some new terms for them. . . and scrub your hands really well when you are done if that is your project.

C. Historical Perspective
When did we first learn about this idea, and who is responsible for getting us this far? You need to give a historical perspective with names, dates, countries, awards, and other recognition.

Building a Research Foundation

1. This sheet is designed to help you organize your thoughts and give you some ideas on where to look for information on your topic. When you prepare your lab report, you will want to include the background information outlined below.

A. *Major Scientific Concepts (Two is plenty.)*

   i. _____

   _____

   _____

   _____

   ii. _____

   _____

   _____

   _____

B. *Scientific Words (No more than 10)*

   i. _____

   ii. _____

   iii. _____

   iv. _____

   v. _____

   vi. _____

   vii. _____

   viii. _____

   ix. _____

   x. _____

C. *Historical Perspective*
   Add this as you find it.

2. There are several sources of information that are available to help you fill in the details from the previous page.

A. *Contemporary Print Resources*
   (Magazines, Newspapers, Journals)

   i. _____
   ii. _____
   iii. _____
   iv. _____
   v. _____
   vi. _____

B. *Other Print Resources*
   (Books, Encyclopedias, Dictionaries, Textbooks)

   i. _____
   ii. _____
   iii. _____
   iv. _____
   v. _____
   vi. _____

C. *Celluloid Resources*
   (Films, Filmstrips, Videos)

   i. _____
   ii. _____
   iii. _____
   iv. _____
   v. _____
   vi. _____

D. *Electronic Resources:*
   (Internet Website Addresses, DVDs, MP3s)

   i. _____

   ii. _____

   iii. _____

   iv. _____

   v. _____

   vi. _____

   vii. _____

   viii. _____

   ix. _____

   x. _____

E. *Human Resources*
   (Scientists, Engineers, Professionals, Professors, Teachers)

   i. _____

   ii. _____

   iii. _____

   iv. _____

   v. _____

   vi. _____

You may want to keep a record of all of your research and add it to the back of the report as an Appendix. Some teachers who are into volume think this is really cool. Others, like myself, find it a pain in the tuchus. No matter what you do, be sure to keep an accurate record of where you find data. If you quote from a report word for word, be sure to give proper credit with either a footnote or parenthetical reference, this is very important for credibility and accuracy. This is will keep you out of trouble with plagiarism (copying without giving credit).

Scientific Method
• Step 3 •
Design Your Experiment

Electron Herding 101 • © 2002 • B. K Hixson

Acquire Your Lab Materials

The purpose of this section is to help you plan your experiment. You'll make a map of where you are going , how you want to get there, and what you will take along.

List the materials you will need to complete your experiment in the table below. Be sure to list multiples if you will need more than one item. Many science materials double as household items in their spare time. Check around the house before you buy anything from a science supply company or hardware store. For your convenience, we have listed some suppliers on page 19 of this book.

Material	Qty.	Source	$
1.			
2.			
3.			
4.			
5.			
6.			
7.			
8.			
9.			
10.			
11.			
12.			

Total $_____

Outline Your Experiment

This sheet is designed to help you outline your experiment. If you need more space, make a copy of this page to finish your outline. When you are done with this sheet, review it with an adult, make any necessary changes, review safety concerns on the next page, prepare your data tables, gather your equipment, and start to experiment.

In the space below, list what you are going to do in the order you are going to do it.

i. _____

_____

_____

_____

ii. _____

_____

_____

_____

iii. _____

_____

_____

_____

iv. _____

_____

_____

_____

v. _____

_____

_____

_____

Evaluate Safety Concerns

We have included an overall safety section in the front of this book on pages 16–18, but there are some very specific questions you need to ask, and prepare for, depending on the needs of your experiment. If you find that you need to prepare for any of these safety concerns, place a check mark next to the letter.

_____ *A. Goggles & Eyewash Station*
If you are mixing chemicals or working with materials that might splinter or produce flying objects, goggles and an eyewash station or sink with running water should be available.

_____ *B. Ventilation*
If you are mixing chemicals that could produce fire, smoke, fumes, or obnoxious odors, you will need to use a vented hood or go outside and perform the experiment in the fresh air.

_____ *C. Fire Blanket or Fire Extinguisher*
If you are working with potentially combustible chemicals or electricity, a fire blanket and extinguisher nearby are a must.

_____ *D. Chemical Disposal*
If your experiment produces a poisonous chemical or there are chemical-filled tissues (as in dissected animals), you may need to make arrangements to dispose of the by-products from your lab.

_____ *E. Electricity*
If you are working with materials and developing an idea that uses electricity, make sure that the wires are in good repair, that the electrical demand does not exceed the capacity of the supply, and that your work area is grounded.

_____ *F. Emergency Phone Numbers*
Look up and record the following phone numbers for the Fire Department: _____ , Poison Control: _____ , and Hospital: _____. Post them in an easy-to-find location.

Prepare Data Tables

Finally, you will want to prepare your data tables and have them ready to go before you start your experiment. Each data table should be easy to understand and easy for you to use.

A good data table has a **title** that describes the information being collected, and it identifies the **variable** and the **unit** being collected on each data line. The variable is *what* you are measuring and the unit is *how* you are measuring it. They are usually written like this:

Variable (unit), or to give you some examples:

> *Time (seconds)*
> *Distance (meters)*
> *Electricity (volts)*

An example of a well-prepared data table looks like the sample below. We've cut the data table into thirds because the book is too small to display the whole line.

Determining the Boiling Point of Compound X_1

Time (min.)	0	1	2	3	4	5	6
Temp. (ºC)							

Time (min.)	7	8	9	10	11	12	13
Temp. (ºC)							

Time (min.)	14	15	16	17	18	19	20
Temp. (ºC)							

Scientific Method
• Step 4 •
Conduct the Experiment

Lab Time

It's time to get going. You've generated a hypothesis, collected the materials, written out the procedure, checked the safety issues, and prepared your data tables. Fire it up. Here's the short list of things to remember as you experiment.

_____ *A. Follow the Procedure, Record Any Changes*

Follow your own directions specifically as you wrote them. If you find the need to change the procedure once you are into the experiment, that's fine; it's part of the process. Make sure to keep detailed records of the changes. When you repeat the experiment a second or third time, follow the new directions exactly.

_____ *B. Observe Safety Rules*

It's easier to complete the lab activity if you are in the lab rather than the emergency room.

_____ *C. Record Data Immediately*

Collect temperatures, distances, voltages, revolutions, and any other variables and immediately record them into your data table. Do not think you will be able to remember them and fill everything in after the lab is completed.

_____ *D. Repeat the Experiment Several Times*

The more data that you collect, the better. It will give you a larger data base and your averages are more meaningful. As you do multiple experiments, be sure to identify each data set by date and time so you can separate them out.

_____ *E. Prepare for Extended Experiments*

Some experiments require days or weeks to complete, particularly those with plants and animals or the growing of crystals. Prepare a safe place for your materials so your experiment can continue undisturbed while you collect the data. Be sure you've allowed enough time for your due date.

Scientific Method
• Step 5 •
Collect and Display Data

Types of Graphs

This section will give you some ideas on how you can display the information you are going to collect as a graph. A graph is simply a picture of the data that you gathered portrayed in a manner that is quick and easy to reference. There are four kinds of graphs described on the next two pages. If you find you need a leg up in the graphing department, we have a book in the series called *Data Tables & Graphing*. It will guide you through the process.

Line and Bar Graphs

These are the most common kinds of graphs. The most consistent variable is plotted on the "x", or horizontal, axis and the more temperamental variable is plotted along the "y", or vertical, axis. Each data point on a line graph is recorded as a dot on the graph and then all of the dots are connected to form a picture of the data. A bar graph starts on the horizontal axis and moves up to the data line.

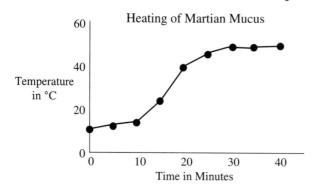

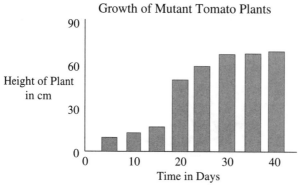

Electron Herding 101 • © 2002 • B. K Hixson

Best Fit Graphs
A best fit graph was created to show averages or trends rather than specific data points. The data that has been collected is plotted on a graph just as on a line graph, but instead of drawing a line from point to point to point, which sometimes is impossible anyway, you just free hand a line that hits "most of the data."

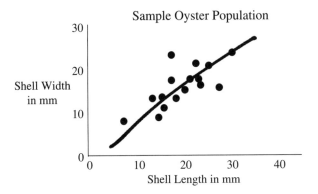

Pie Graphs
Pie graphs are used to show relationships between different groups. All of the data is totaled up and a percentage is determined for each group. The pie is then divided to show the relationship of one group to another.

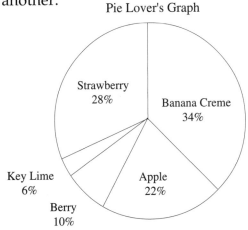

Other Kinds of Data

1. Written Notes & Observations

This is the age-old technique used by all scientists. Record your observations in a lab book. Written notes can be made quickly as the experiment is proceeding, and they can then be expounded upon later. Quite often notes made in the heat of an experiment are revisited during the evaluation portion of the process, and they can shed valuable light on how or why the experiment went the way it did.

2. Drawings

Quick sketches as well as fully developed drawings can be used as a way to report data for a science experiment. Be sure to title each drawing and, if possible, label what it is that you are looking at. Drawings that are actual size are best.

3. Photographs, Videotapes, and Audiotapes

Usually better than drawings, quicker, and more accurate, but you do have the added expense and time of developing the film. However, they can often capture images and details that are not usually seen by the naked eye.

4. The Experiment Itself

Some of the best data you can collect and present is the actual experiment itself. Nothing will speak more effectively for you than the plants you grew, the specimens you collected, or that big pile of tissue that was an armadillo you peeled from the tread of an 18-wheeler.

Scientific Method
· Step 6 ·
Present Your Ideas

Oral Report Checklist

It is entirely possible that you will be asked to make an oral presentation to your classmates. This will give you an opportunity to explain what you did and how you did it. Quite often this presentation is part of your overall score, so if you do well, it will enhance your chances for one of the bigger awards.

To prepare for your oral report, your science fair presentation should include the following components:

Physical Display

_____a.   freestanding display board
                    hypothesis
                    data tables, graphs, photos, etc.
                    abstract (short summary)
_____b.   actual lab setup (equipment)

Oral Report

_____a.   hypothesis or question
_____b.   background information
                    concepts
                    word definitions
                    history or scientists
_____c.   experimental procedure
_____d.   data collected
                    data tables
                    graphs
                    photos or drawings
_____e.   conclusions and findings
_____f.   ask for questions

Set the display board up next to you on the table. Transfer the essential information to index cards. Use the index cards for reference, but do not read from them. Speak in a clear voice, hold your head up, and make eye contact with your peers. Ask if there are any questions before you finish and sit down.

Written Report Checklist

Next up is the written report, also called your lab write-up. After you compile or sort the data you have collected during the experiment and evaluate the results, you will be able to come to a conclusion about your hypothesis. Remember, disproving an idea is as valuable as proving it.

This sheet is designed to help you write up your science fair project and present your data in an organized manner. This is a final checklist for you.

To prepare your write-up, your science fair report should include the following components:

_____ a.   binder
_____ b.   cover page, title, & your name
_____ c.   abstract (one paragraph summary)
_____ d.   table of contents with page numbers
_____ e.   hypothesis or question
_____ f.   background information
                     concepts
                     word definitions
                     history or scientists
_____ g.   list of materials used
_____ h.   experimental procedure
                     written description
                     photo or drawing of setup
_____ i.   data collected
                     data tables
                     graphs
                     photos or drawings
_____ j.   conclusions and findings
_____ k.   glossary of terms
_____ l.   references

Display Checklist

2. Prepare your display to accompany the report. A good display should include the following:

Freestanding Display

_____ a.   freestanding cardboard back
_____ b.   title of experiment
_____ c.   your name
_____ d.   hypothesis
_____ e.   findings of the experiment
_____ f.   photo or illustrations of equipment
_____ g.   data tables or graphs

Additional Display Items

_____ h.   a copy of the write-up
_____ i.   actual lab equipment setup

Glossary,
Index,
and
More Ideas

Glossary

Atomic Model
Model of atoms depicting protons, neutrons, and electrons in correct ratio to one another—generally pictured as round balls hooked together.

Battery
An electrical storage device. Can consist of two different metals in a conducting liquid, two metals heated together, or panels that convert the energy of the sun to electrical energy.

Charge
The opposite of retreat. But when studying electricity, one tends to think of electrons, which, when they accumulate, have a negative charge. In their absence, a positive charge is created.

Circuit
A complete loop made of conductive material and connected to a source of electrons.

Compass
An instrument consisting of a magnetized needle suspended and spinning freely on a needle. The magnetized part of the instrument is influenced by and responds to the Earth's magnetic field, and the end of the needle points toward magnetic north.

Conduction
The ability of a material to allow electrons to pass through it.

Conductors
Materials that allow electricity to flow through them freely—typically metals and liquids that have ionized particles.

Copper
An element in the Periodic Table and a principal component of most batteries.

Dimmer
An instrument that takes advantage of some materials that resist but do not prevent the movement of electrons. When used to increase or decrease the amount of light in a lamp, they are called dimmer switches; and when they are rotated, they either increase or decrease the flow of electrons to the lamp.

Discharge
Releasing electrons that have accumulated in or on a material in large quantity and are tired of being crammed together, so they are anxious to jump to a new location and spread out.

Electromagnets
Magnets created when electricity flows through a conductive material, like copper wire, around an iron core.

Electrons
Small, negatively charged particles that are associated with atomic nuclei. They are very small particles and can be swiped very easily from the atoms.

Electrophorus
An insulated plastic or vinyl sheet that is rubbed on an electron-rich source such as flannel or wool. The vinyl picks up large amounts of electrons creating a huge negative charge that can then be collected to a Leyden jar or discharged on a discharge rod.

Electroplating
Using an electric current in a conductive liquid to move charged metal ions onto a metal surface.

Glossary

Electroscope

A simple device used to detect electric charge. The charge is collected at the top of the electroscope and distributed down to small moveable objects like foil leaves or rice puffs. When they repel one another, it indicates a negative charge.

Flashlight

An instrument that converts electrical energy into light on command.

Fuses

A calibrated piece of wire that is specifically designed to only carry a certain amount of electrical energy. If the amount of electricity trying to flow through the wire exceeds the capacity of the fuse, it melts and disconnects the circuit. Fuses are used to protect expensive equipment from catching on fire or melting.

Galvanometer

A coil of wire looped around a compass that detects weak electromagnetic fields that can be induced in wires by magnets.

Heater

An instrument that converts electrical energy into heat.

Induced Current

When a permanent magnet is inserted into a coil of wire, it causes an electrical current to begin to flow in the coil. This is said to be an induced current.

Insulators

Materials that do not allow electricity to flow through them.

Leyden Jar

A primitive storage device for electrons, consisting of a glass jar wrapped in aluminum. When charged with static electricity, it can deliver a good zap when brought near a conductor.

Lightning Rod

A metal rod that is placed in a high, conspicuous place specifically for the purpose of attracting lightning bolts so that they go there rather than fry the cow grazing in the pasture or zap the postman on his local rounds.

Morse Code

A series of dots and dashes used to communicate over the telegraph lines that was used in the mid 1800s. Designed by Samuel Morse it was replaced by the invention of the telephone.

Motor Effect

The movement or a wire or coil of wires when an electric current is applied to a wire between two magnets. It is the push that gets the coil moving in a real motor.

Neutrons

Large particles found in the center or nucleus of the atom. They do not have a charge.

Nichrome Wire

A metal alloy made of nickel and chromium. This particular metal resists the movement of electrons through it and some of the energy is used to produce heat and light. Your toaster and toaster oven are made using nichrome wires.

Nucleus

This is the home base of the atom consisting of protons, positively charged particles, and neutrons, particles that have no real opinion on whether they should be charged at all, so they are not. The nucleus of the atom is responsible for 99.9% of the mass of the atom and is also how the atomic number is calculated. *Nuclei* is the plural of *nucleus*, a term that originated in Utah in the late 1800s.

Glossary

Phosphor
A white compound that is used to coat the inside of fluorescent lightbulbs. When the bulbs are plugged in, the electrical elements excite gases trapped inside the glass bulbs. When the gases are excited, they emit electrons that escape through the glass. While they are escaping, they pass through the phosphor and the electrons excite the phosphor, which responds by emitting light.

Piezoelectric Rocks
Rocks that produce light when they are smashed together.

Protons
Large, positively charged particles that comprise the center of the atom with neutrons.

Resistance
The ability of a material to prevent electrons from passing through it—typically metals and liquids that have ionized particles.

Schematic Symbols
Shorthand for electricians. Instead of drawing a picture or writing a word to represent bulbs, lights, wires, batteries, speakers, motors, and other assorted electrical hoo hah, they use simple symbols that can be easily inserted into a drawing. A wire, for example, is a line.

Speaker
An instrument that convert electrical energy into sound.

Static Electricity
An electrical charge that is accumulated and transferred in a random, unpredictable, and measurable way. Generally, rubbing a plastic or vinyl substance with flannel will gather and collect a large pile of electrons. These electrons are called static electricity, because it is electricity, but you can't control it or get it to move from one place to another very easily.

Switches
An instrument that controls the flow of electricity in an electrical circuit.

Van de Graaff Generator
One of the greatest science toys ever invented. A belt spins around a felt roller. As it spins up into a metal dome, the electrons that are collected by the rubber belt are discharged onto the dome and accumulate. When a conductive material, like a finger, comes in contact with the dome, the electrons jump en masse to the finger and a spark of light, a loud snap, and a startled look can all be seen.

Vinegar
A common electrolyte.

Zinc
An element in the Periodic Table and a principal component of most batteries.

Index

Index

Notes

Notes

Notes

More Physics Books

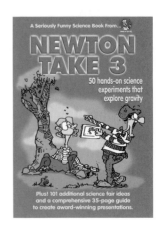

Catch a Wave
40 hands-on lab activities that sound off on the topic of noise, vibration, waves, the Doppler Effect and associated ideas.

Thermodynamic Thrills
35 hands-on lab activities that investigate heat via conduction, convection, radiation, specific heat, and temperature.

Newton Take 3
50 hands-on lab activities that explore the world of mechanics, forces, gravity, and Newton's three laws of motion.

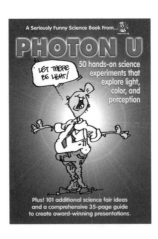

Gravity Works
50 hands-on lab activities from the world of things that fly. Air, air pressure, Bernoulli's law, and all things that fly, float, or glide are explored.

Photon U
50 hands-on lab activities that introduce light, the electromagnetic spectrum and include a number of fun, and very easy-to-build projects.

Opposites Attract
35 hands-on lab activities that delve into the world of natural and manmade magnets as well as the characteristics of magnetic attraction.